ADEL ABDESSEMED
FRANZ ACKERMANN
ANGELA BULLOCH
[illegible]
[illegible]
JANET CARDIFF & GEORGE BURES MILLER
CERITH WYN EVANS
[illegible]
[illegible]
THOMAS HIRSCHHORN
[illegible]
KIMSOOJA
[illegible]
[illegible]
JOSIAH McELHENY
[illegible]
GABRIEL OROZCO
THE OTOLITH GROUP
ADRIAN PACI
TREVOR PAGLEN
KATIE PATERSON
FRED TOMASELLI
KEITH TYSON

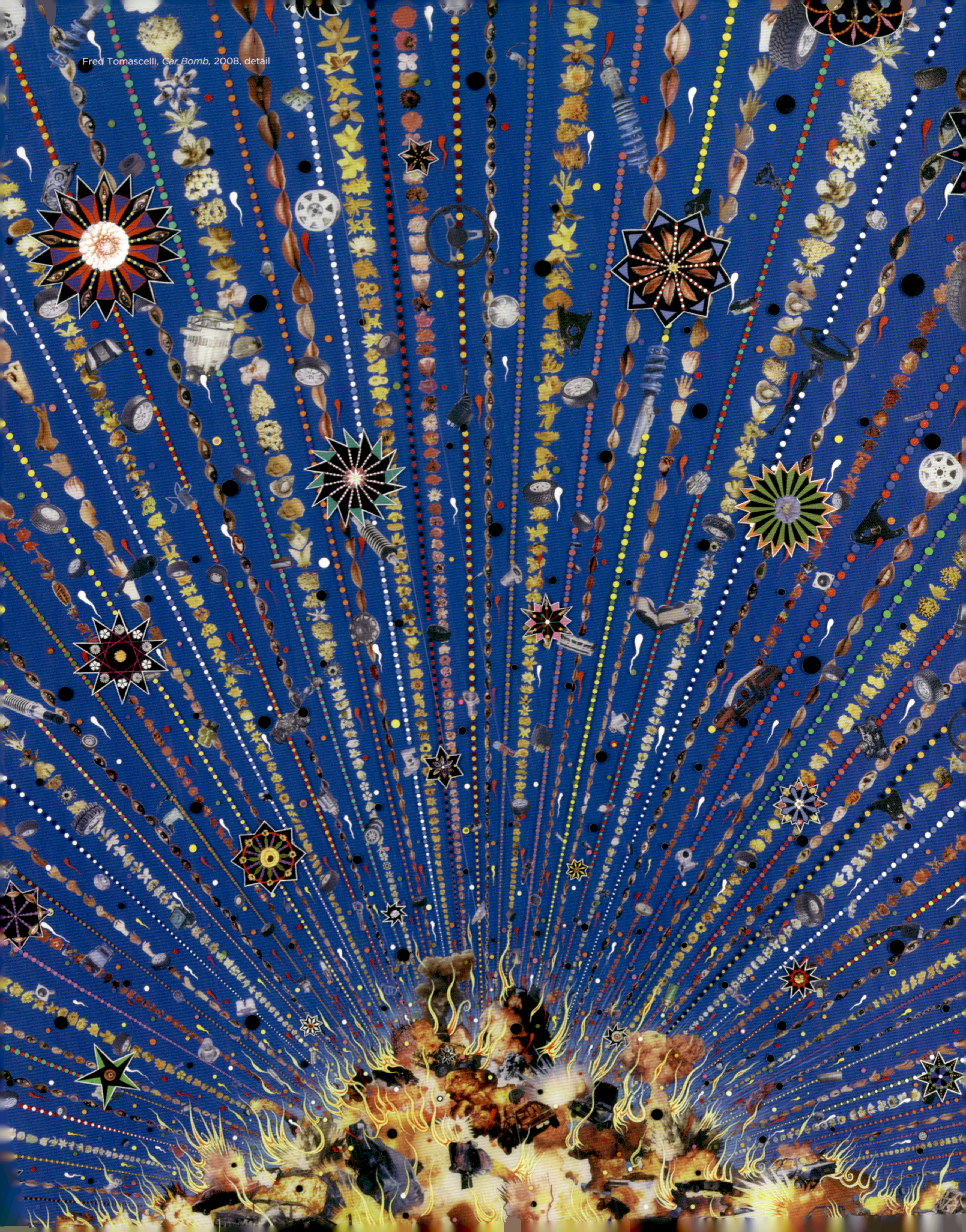
Fred Tomascelli, *Car Bomb*, 2008, detail

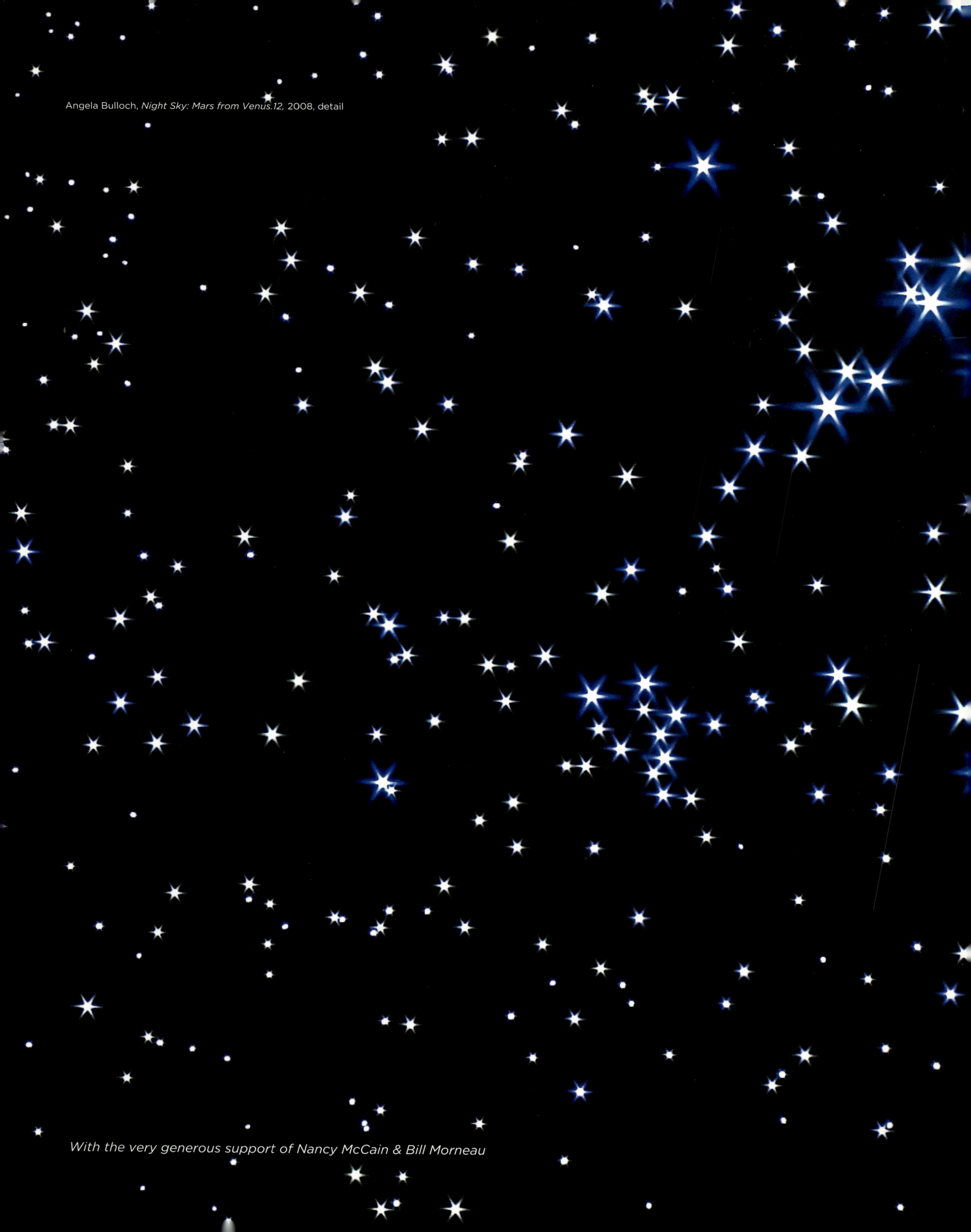

Angela Bulloch, *Night Sky: Mars from Venus.12*, 2008, detail

With the very generous support of Nancy McCain & Bill Morneau

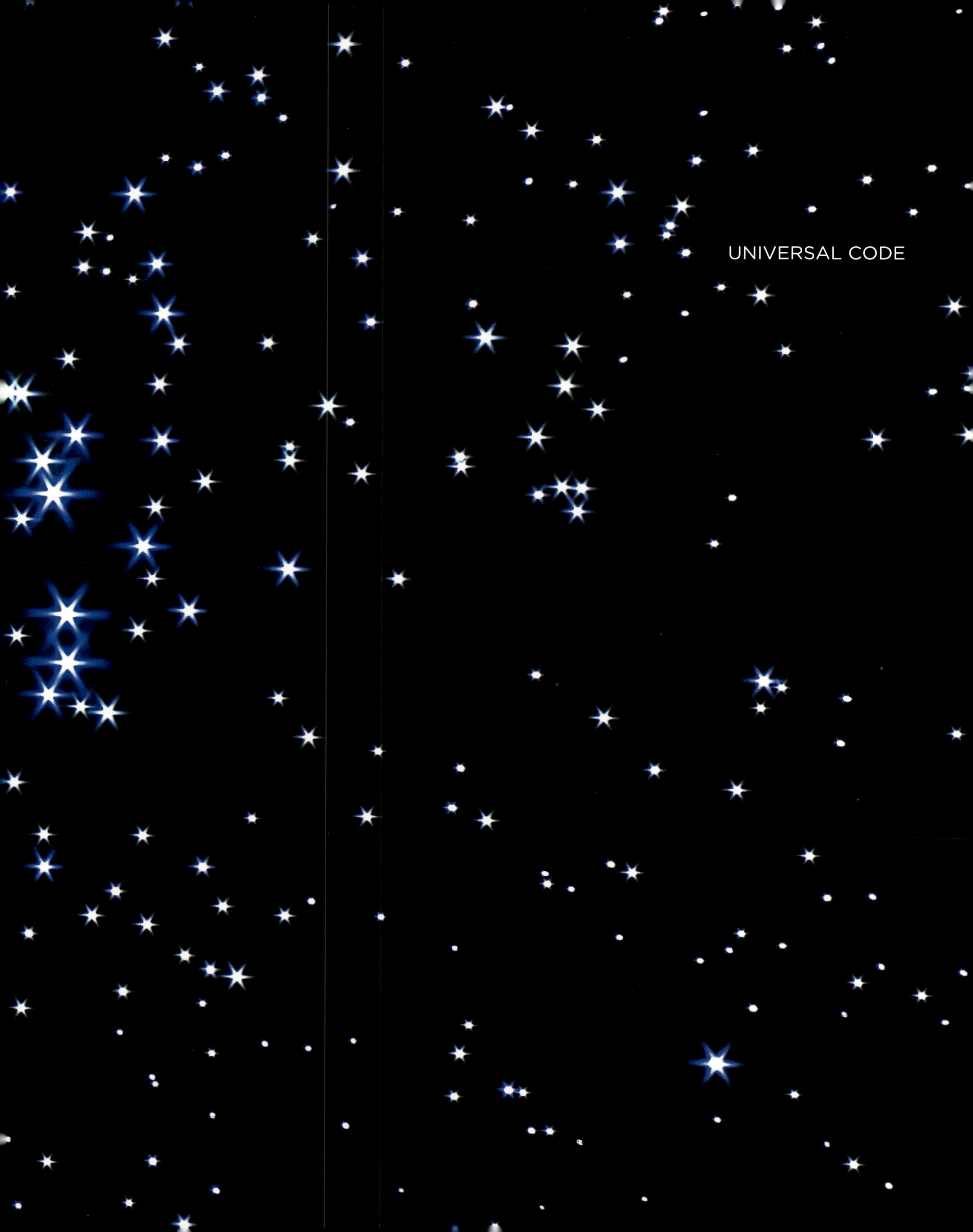

UNIVERSAL CODE

Henrich Håkansson, *Monarch–The Eternal*, 2008, still

Also supported by Elisa Nuyten & David Dime, Laura Rapp & Jay Smith and Garnet & Evan Siddall

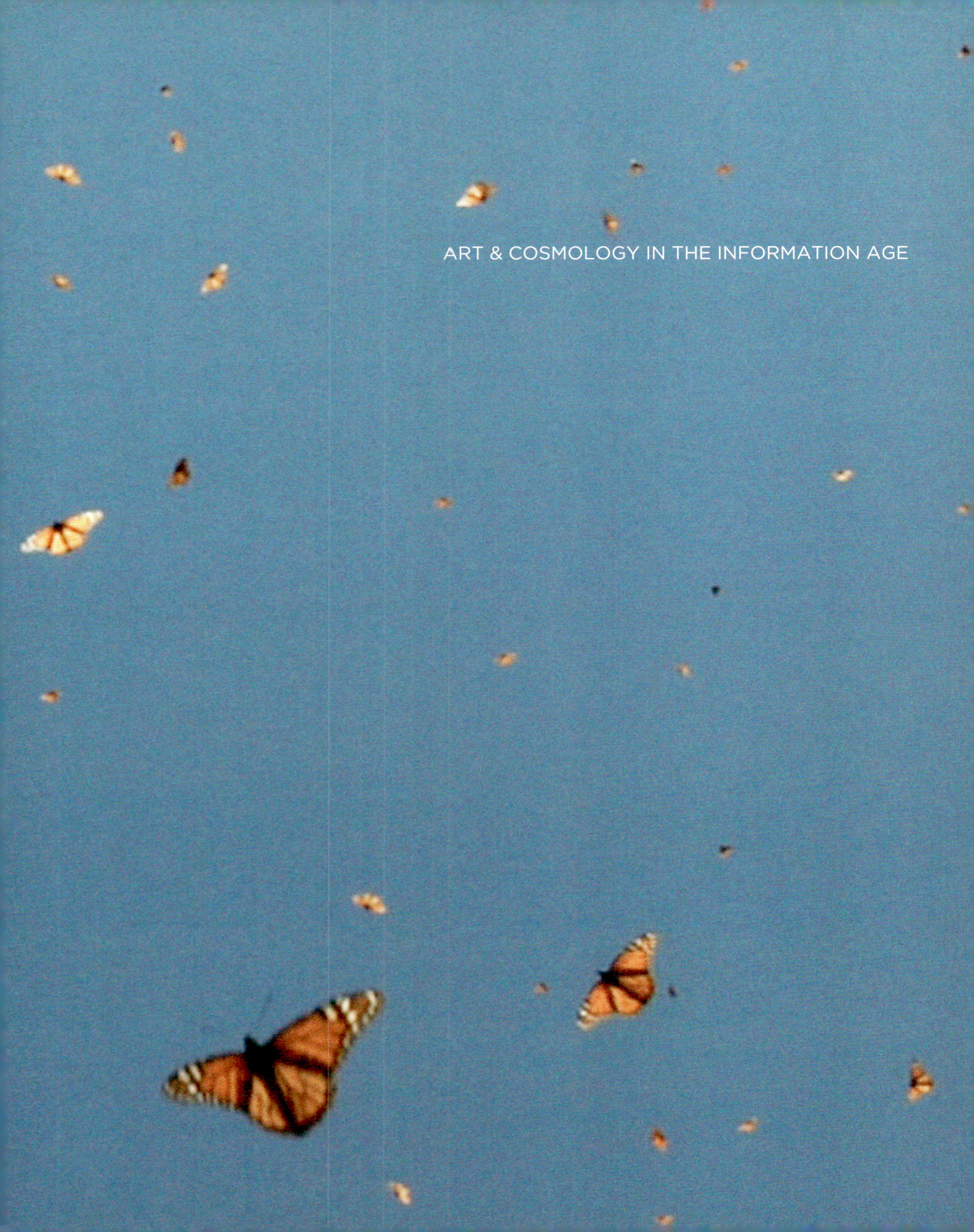

ART & COSMOLOGY IN THE INFORMATION AGE

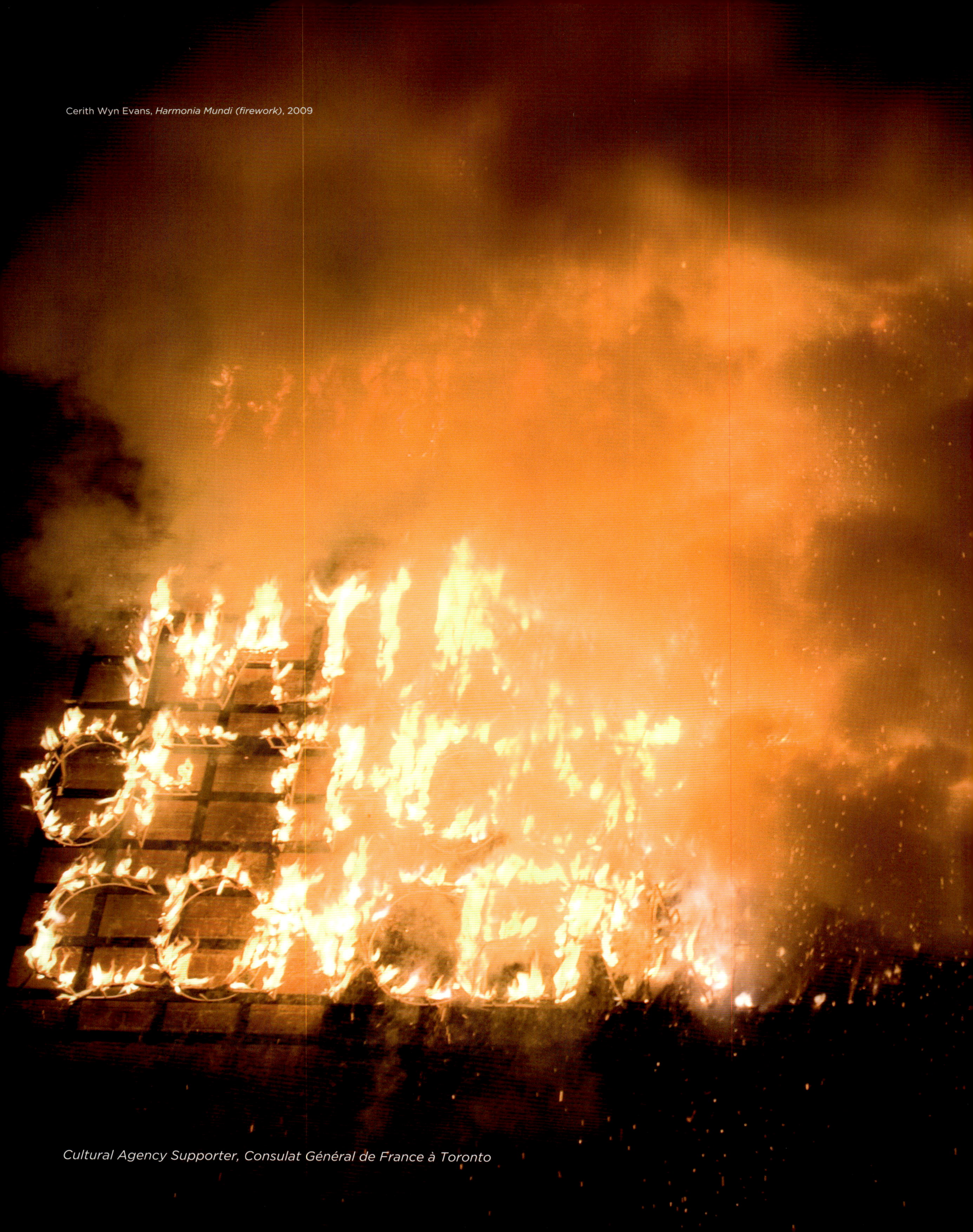

Cerith Wyn Evans, *Harmonia Mundi (firework)*, 2009

Cultural Agency Supporter, Consulat Général de France à Toronto

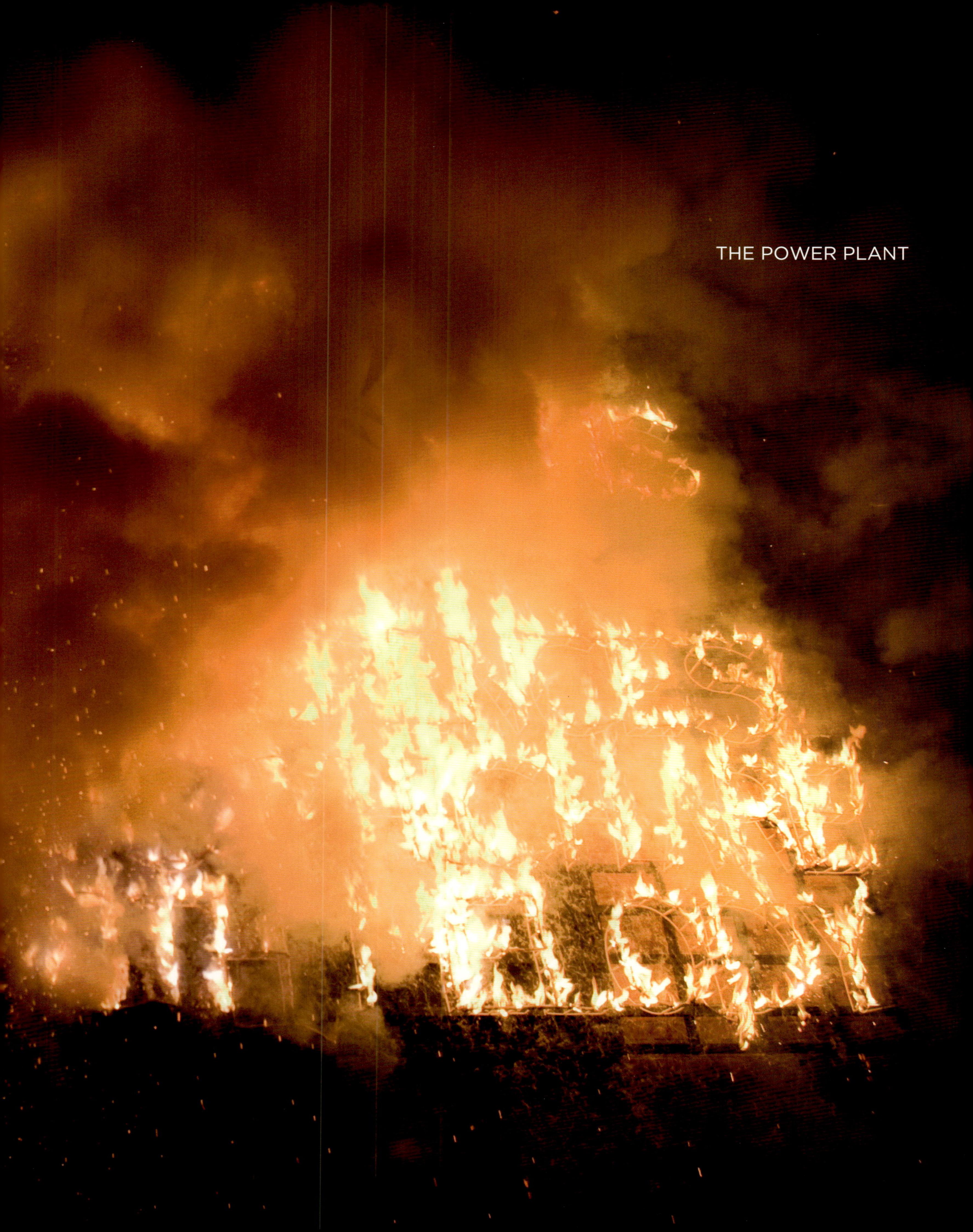

THE POWER PLANT

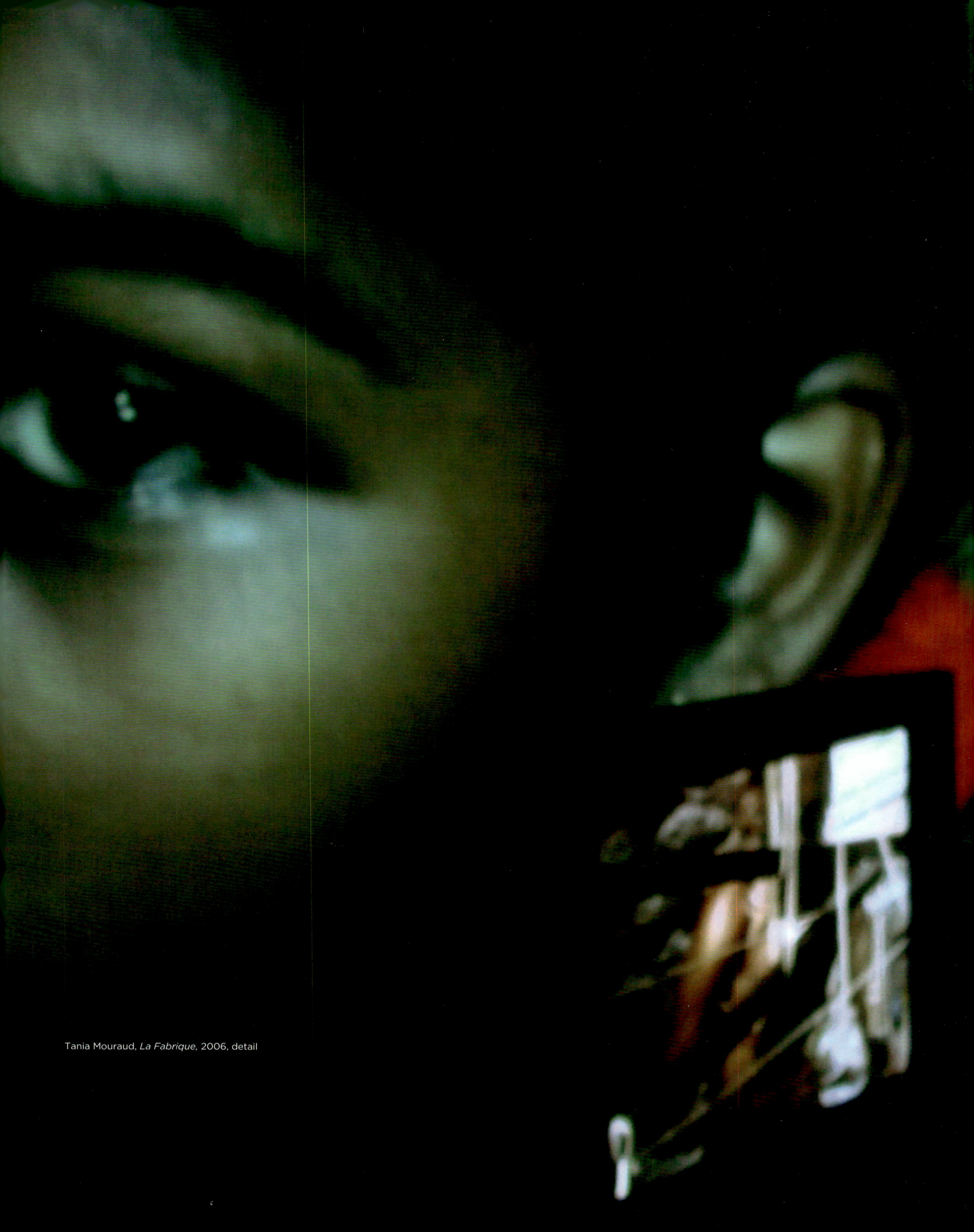

Tania Mouraud, *La Fabrique*, 2006, detail

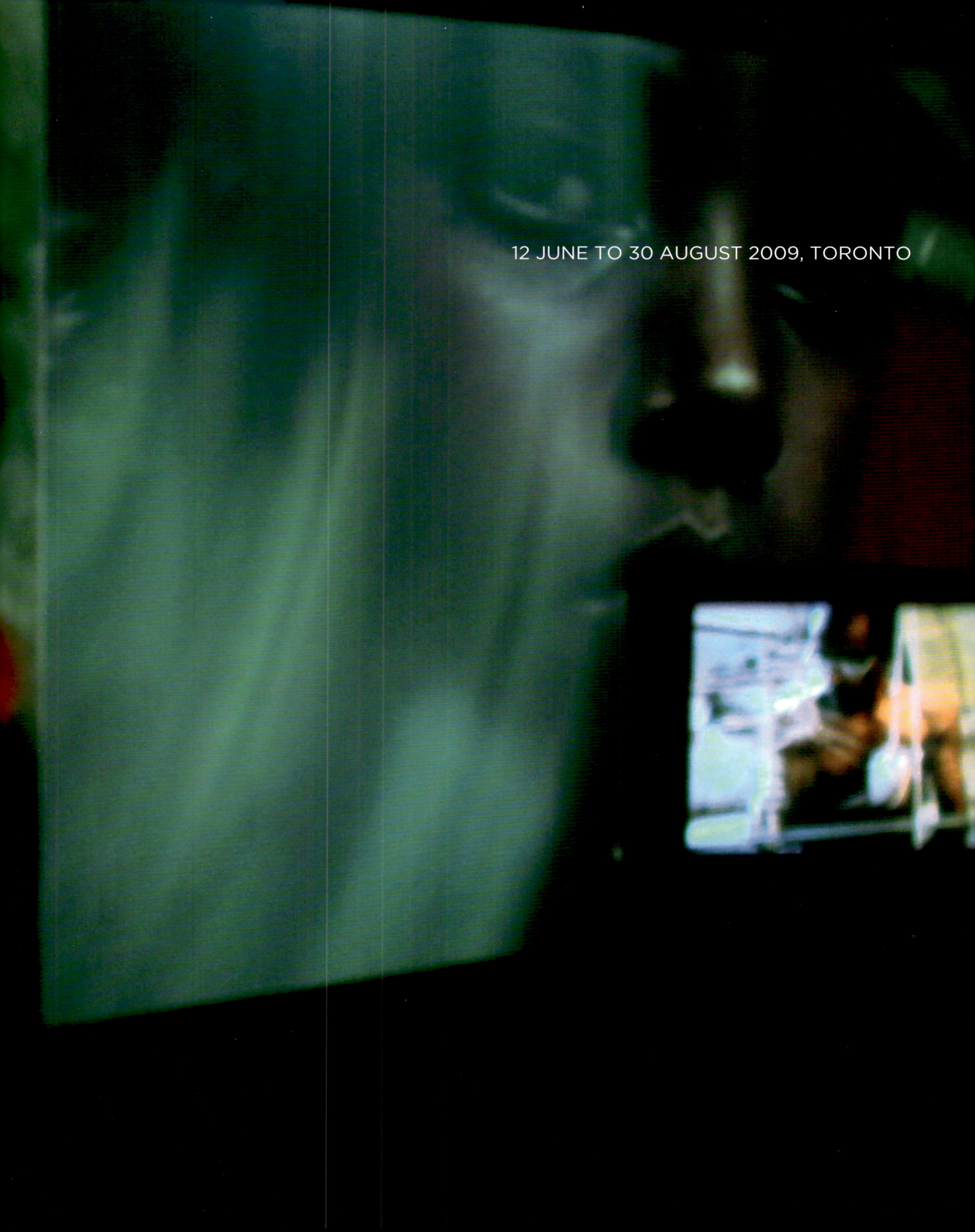

12 JUNE TO 30 AUGUST 2009, TORONTO

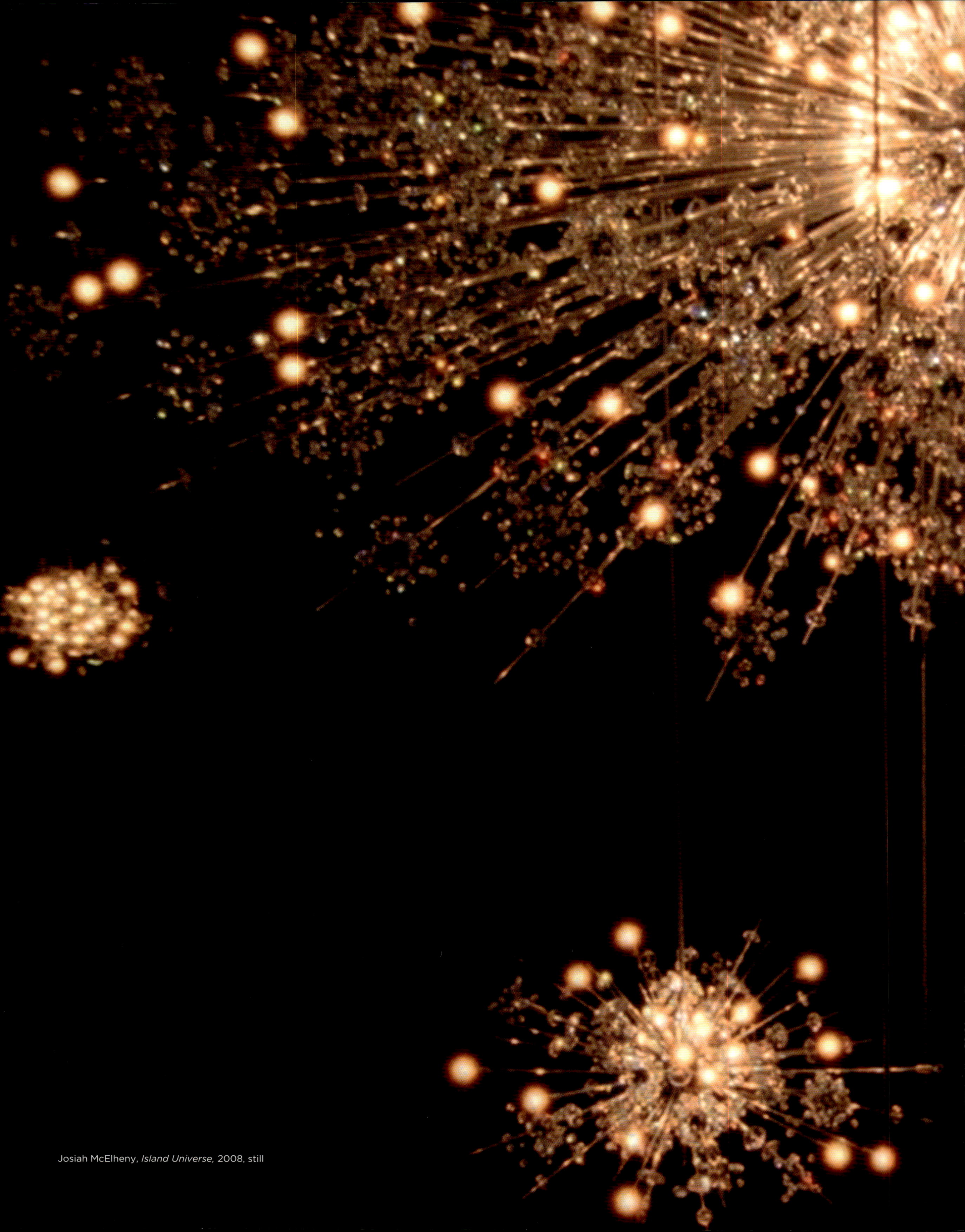

Josiah McElheny, *Island Universe,* 2008, still

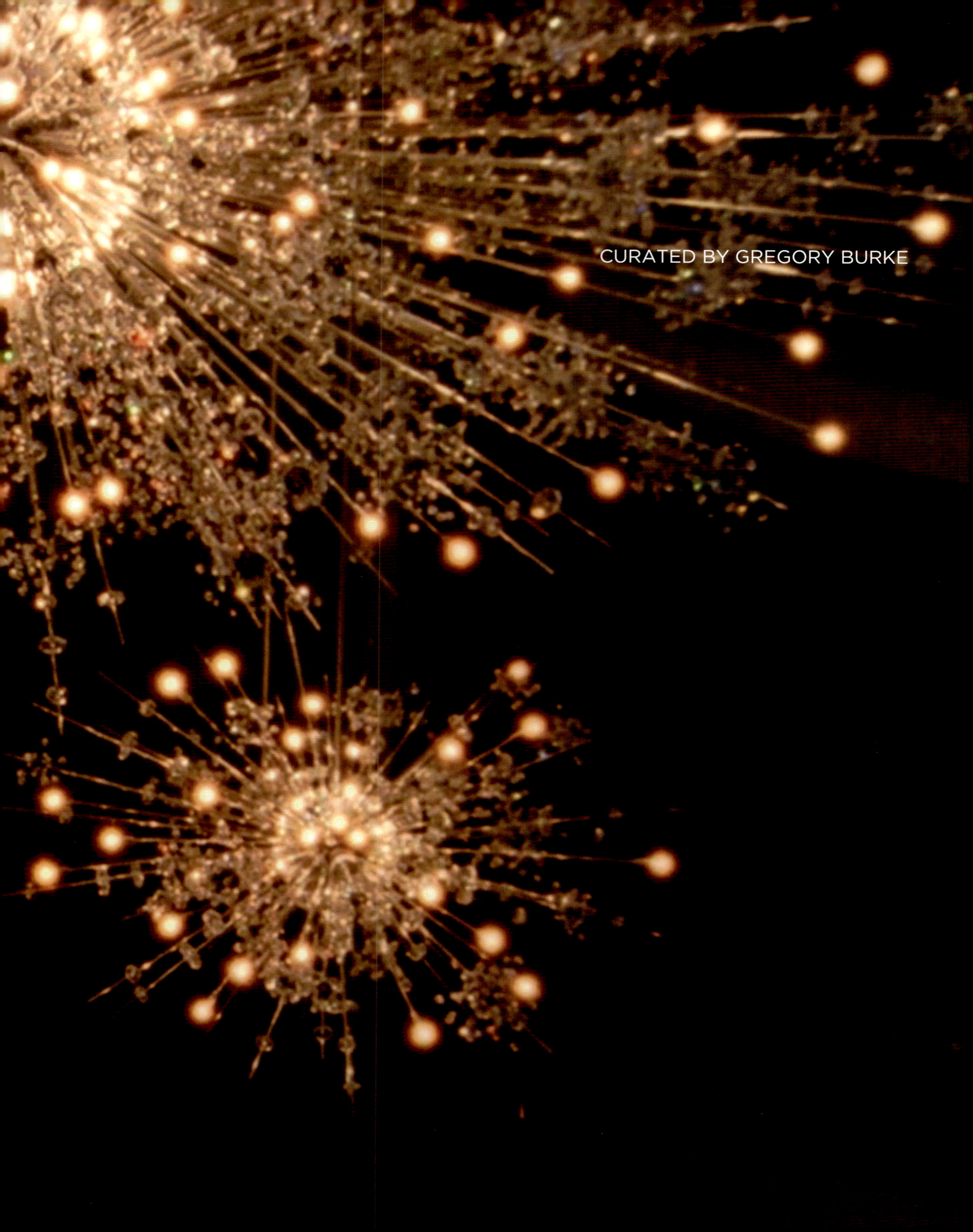

CURATED BY GREGORY BURKE

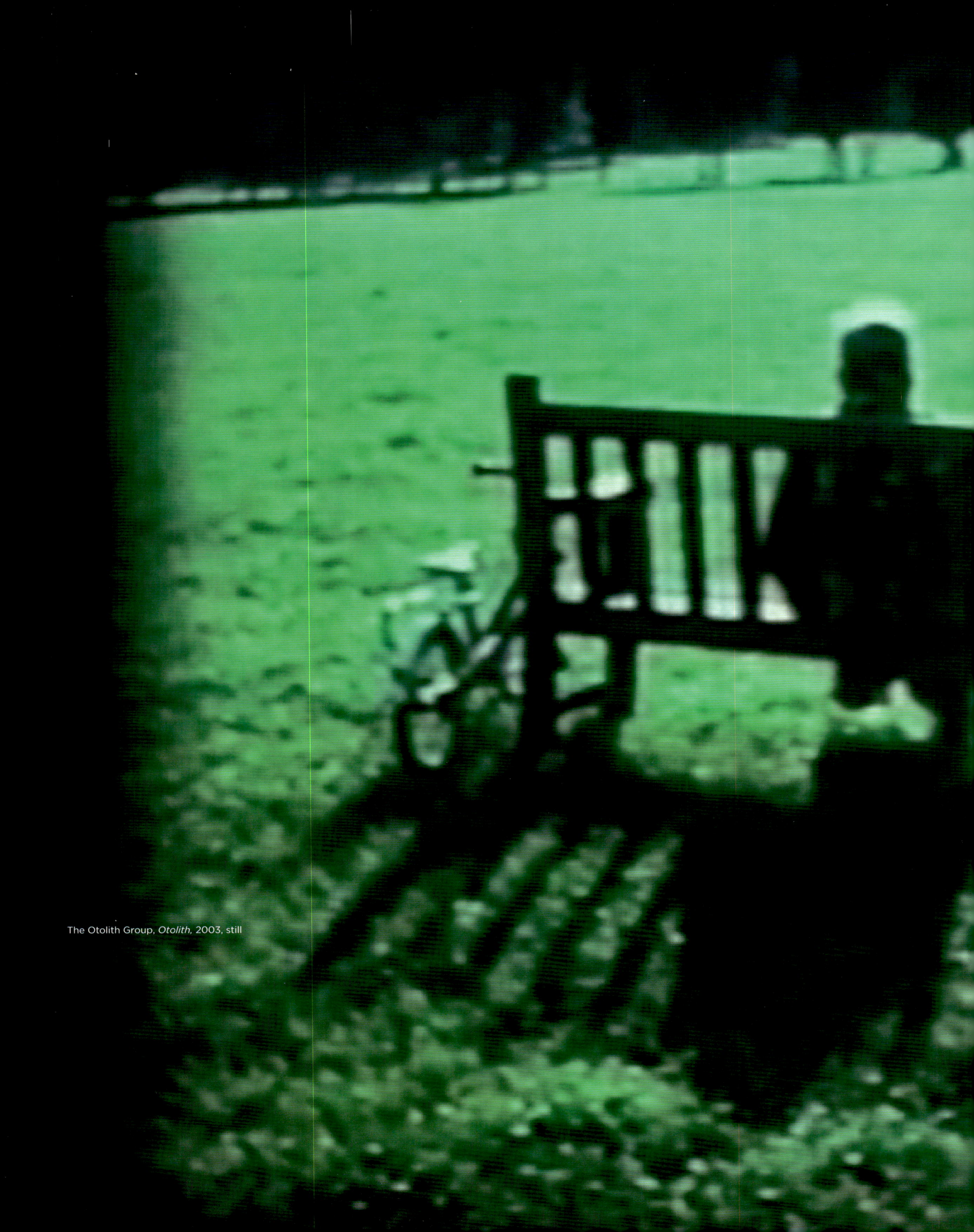

The Otolith Group, *Otolith*, 2003, still

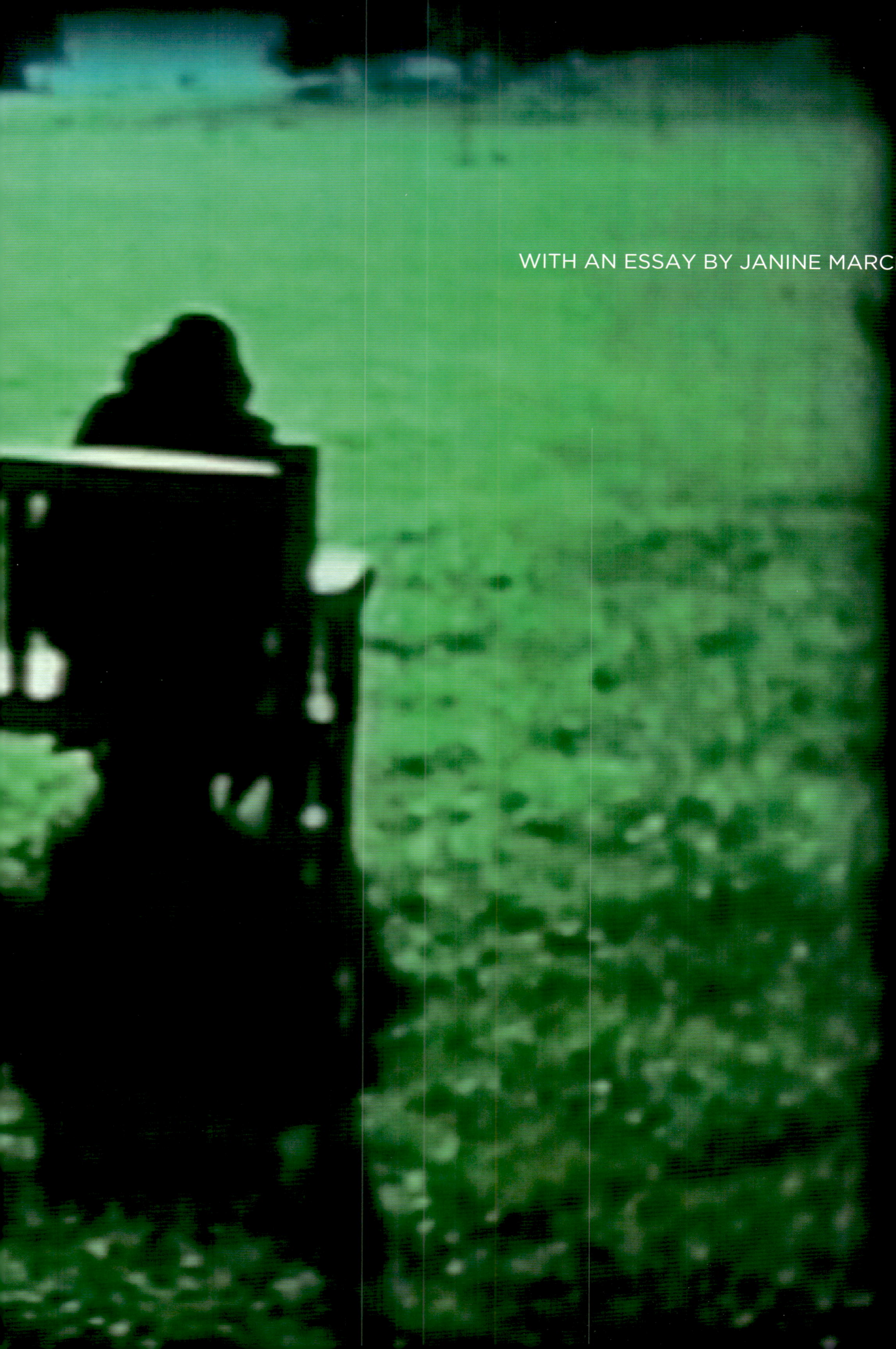

WITH AN ESSAY BY JANINE MARCHESSAULT

CONTENTS

Front

Back

The Exhibition

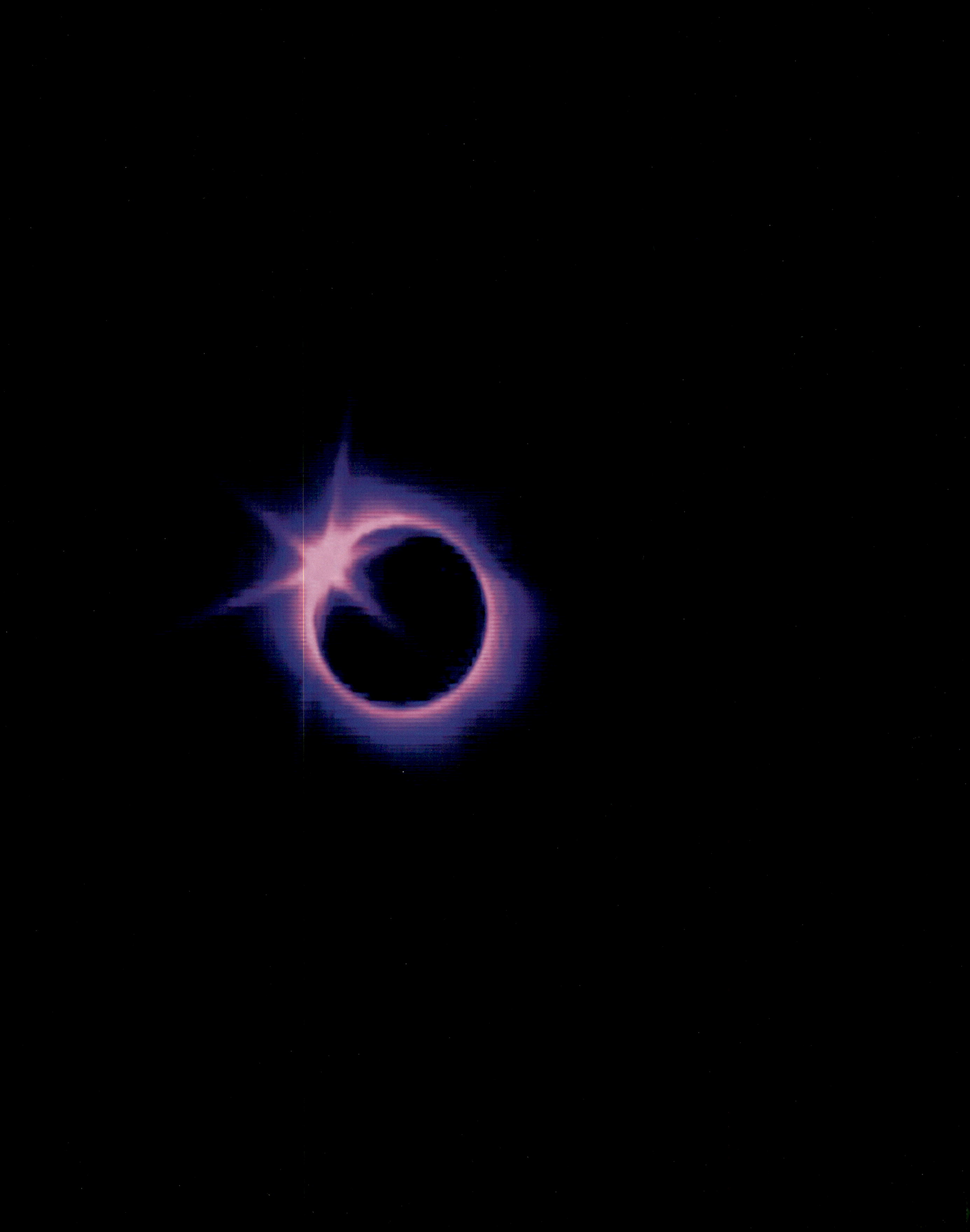

UNIVERSAL CODE
GREGORY BURKE

The cosmos was thus the foundation of the house and the condition for knowledge. Order reigns, all quiet, our science and our confidence.

It is not so: our earth is quaking. The cosmos, the universe, in our languages, are misnomers. It is better seen, ultimately, as an intermediary, it is a mix of order and disorder.[1]

'UNIVERSAL CODE' IS THE LATEST IN A SERIES OF EXHIBITIONS ORGANIZED by The Power Plant, which bring together international artists from a variety of cultural and philosophical positions to reflect on topics driving the development of contemporary culture. 'Universal Code' presents responses from a broad range of contemporary artists to cosmology and ideas of the universal in our current information age, a time of significant scientific and technological development as well as rampant globalization.

Whereas once the sky suggested the limitless, now a network of satellites enmesh the globe in a massive field of signals that unites the world in a "universal" web of telecommunications. Enabling instantaneous connections across vast distances, the new technologies also render visible what was formerly hidden at the bodily and even the cellular level. Corporate communications networks and digital technologies have in effect refigured the contemporary sense of the world and the universe by transforming historical understandings of time and space. In response to this current condition, 'Universal Code' presents artists who look back on the modern era and beyond to consider the universe, the infinite and the eternal from a range of differing religious, scientific, philosophical, and cultural positions.

Works in the exhibition draw and reflect on scientific discoveries related to the universe and the nature of life, from the mapping of galactic space through to research into the microcosmic world of DNA coding. However, the exhibition is not a celebration of scientific and technological advance. While many of the artists seek to engage the viewer in the awe and enigma of concepts of the universe, they are also concerned with shifts in the geopolitical order brought on by the digital revolution, coupled with unresolved tensions that exist between differing belief systems.

In looking back to former times, such as the beginnings of the scientific revolution or the pioneering era of space exploration, many of the artists strive to elucidate aspects of the human condition in the current moment. Others address the increasing levels of surveillance that have developed in recent times and the incursion of satellite tracking systems into the private life of the individual. Collectively they re-examine the idea that the digital era has ushered in a space of total communication. They ponder the disconnection that persists between differing belief systems, thereby questioning the myth of a globalized and, by implication, connected world. Ultimately they point to new ways of thinking about the universe and its myriad possibilities for connection.

ANN VERONICA JANSSENS
Side (Studio Version), 2006, still/detail
Video projection, 3 min., 3 sec., loop, colour, silent.
Courtesy Galerie Micheline Szwajcer, Antwerp

TEMPORALITY/CONNECTIVITY

> *Time is a chaos at first, it is first of all a disorder and a noise . . .*
> *The present now is this indiscernible jumble.*[2]

> *The taste for collecting – in this case codes – does not disappear from this "new astronomy," even if it seeks to be based entirely on "physical causes." Kepler's universe, entirely subject to "physical" laws, is also an immense collection of "metaphysical" curiosities.*[3]

'Universal Code' opened with a sculptural event: the fireworks text *Harmonia Mundi (firework)*, 2009, by Cerith Wyn Evans performed a quotation from German mathematician and astronomer Johannes Kepler. In his work Kepler sought to reconcile divisions in the seventeenth century between new theories of the universe brought on by the scientific revolution and the traditional beliefs of the church. Importantly his celestial physics included a mystical dimension. The Kepler text enacted by Wyn Evans alluded to differences being a fundamental condition of an all-embracing and numinous system of connections. Wyn Evans's exploration of connective systems in Kepler's text revolved around temporality. The work itself lasted exactly one minute; referencing an influential text from three centuries ago, which itself owed a debt to the metaphysics of Aristotle, its method involving the ancient technology of fireworks.

The exploration of the relationship of connectivity to temporality is a thread that weaves through 'Universal Code.' Many of the works are durational, involving sequential audio and visual elements or an aspect of performance. As with the work of Wyn Evans, many also enact a form of dialogue with key aspects of history going back hundreds and sometimes thousands of years. In his 2001–02 work *The Block (Part of Seven Wonders of the World series)*, Keith Tyson takes on the known history of the universe according to the partisan narrative of Western civilization, starting with the Big Bang. In his work *With the Advent of Radio Astronomy . . .*, 2004, Wyn Evans references the development of radio astronomy in the 1960s, a time when the Big Bang theory was being proposed and a moment that ushered in the information age.

With *Transmission (John Cage)*, 2003, Wyn Evans transmits Cage's 1950s "Lecture on Nothing (Silence)" in Morse code. Developed in the nineteenth century during a period of technological advance, yet rendered obsolete by the close of the twentieth century, Morse code has sometimes been described as the first universal code. Encrypted as a series of long and short flashes of light from a lampshade followed by pauses, Wyn Evans's work multiplies the layers of syncopation of Cage's original lecture/composition. In a powerful piece that evokes communication and loss on a cosmic scale, Katie Paterson also uses recomposition and Morse code as part of a process that alters a performance of Beethoven's "Moonlight

Sonata" of 1301. In his film *Island Universe,* 2008, Josiah McElheny goes back to 1965, the year the Lobmeyr chandeliers, which depict differing states of the universe, were designed for the New York Metropolitan Opera. With *Otolith*, 2003, The Otolith Group also explores scientific as well as political developments of the 1950s and 1960s in relation to both futurity and the recent past. Janet Cardiff and George Bures Miller also reference the 1960s in *Telephone/Time*, 2004, principally through furniture and an outdated analog telephone that relays a conversation between Cardiff and a scientist on the nature of time.

In *Dancing Philosophy*, 2002, Thomas Hirschhorn references four philosophers – Antonio Gramsci, Georges Bataille and Gilles Deleuze, who represent significant moments in the course of theoretical thought in the twentieth century, and Baruch Spinoza from the seventeenth century, who Deleuze recognized as a major influence on his own work. Hirschhorn's dance with philosophy is also a dance with time that is echoed in numerous other ways in the exhibition. In particular, the history of modernism, in its most expansive sense (and here I am looking back to the beginning of the scientific revolution), remains a dominant reference point, which raises the question of where the artists may position their own practice in relation to that history. In the closing decades of the twentieth century, many artists implicitly dismantled the meta-narratives of universalism in favour of highlighting difference. At stake were norms that governed the understanding of class, race, sexuality and gender, and in general terms the Eurocentrism of modernist principles of universalism. Such deconstructive tendencies in artistic practice were frequently cited to support the argument that modernism itself had reached a crossroads and that the world had entered an era that came to be described as postmodernity. 'Universal Code' suggests that artists are rethinking modernity by continuing to examine culture in relation to connectivity, but without positioning the universal as a culturally neutral or otherwise pure construct.

Many works in the exhibition also suggest that artists are reconsidering the impacts of modernism by addressing decisive moments in its history. While this may suggest the continuing resonance of modernity within current art practice, it does not necessarily follow that artists are engaged in reaffirming generic understandings of modernist practices. It may be that formative moments in the modern era are referenced to elucidate aspects of our current condition and the contingency of the notion of connectivity. The dilemma here is that in pitching for the universal, one aspiration of modernist art practice was to isolate a pure aesthetic language free of contingency. As Robert Smithson once noted, "pure art tends to view abstraction as independent of nature. *Abstraction rules in a void, pretending to be free of time*."[4] This pretense is clearly not an attribute of the positions taken up in 'Universal Code,' yet there is much about modernism that interests and indeed captivates the artists in the exhibition.

It is also true that the history of modernism revolves around many revisions and philosophical oppositions. As Smithson's remark reveals, aspirations for purity did not reign uncontested. As with artists from any generation, individual positions in 'Universal Code' differ. As a group, though, the artists neither emphatically situate themselves within, nor declare their separation from, the modern, yet they clearly register that aftershocks of the scientific revolution continue into the present. Furthermore, their aim in referencing key thinkers or moments from the past is not ironic quotation but rather re-examination and, in some cases, revitalization. For example, Wyn Evans's *Transmission* is a renewal of Cage's work. It allows us to reconsider Cage's work, free of the frame of time as history. It is one of many cases in the exhibition where ideas or significance can be seen to jump generations. Ultimately the consideration of temporality by the artists in the exhibition is decidedly equivocal. The scientific revolution led to a philosophical debate on temporality: one view being that time is a fundamental dimension of the universe, and as such is sequential, can be observed empirically and is measurable; the other being that time is subjective, a state of mind, and therefore immeasurable. The artists in 'Universal Code' eschew such dialectical arguments and revel instead in the possibility that different modalities of time and systems of connectivity coexist, intermingling in a paradoxical zone of the indecipherable.

ORDER/DISORDER

> *On the whole I would say entropy contradicts the usual notion of a mechanistic world view. In other words it's a condition that's irreversible, it's a condition that's moving towards a gradual equilibrium and it's suggested in many ways.*[5]

> *Time-as-weather, fire-as-heat are not concepts in the usual sense. Clock-time is such a concept, entropic time is, perhaps, another such, and temperature as well. A cloud is an aggregate, a nebulous set, a multiplicity whose exact definition escapes us, and whose local movements are beyond observation.*[6]

How the world is perceived and enacted upon is increasingly dominated by information technologies, a consideration highlighted when entering the 'Universal Code' exhibition, where the first work encountered is *Wind Shear*, 1997, by Iñigo Manglano-Ovalle. One gazes upwards at two surveillance-style video monitors showing the movement of clouds across the sky, the jumping-off point to the universe. The work involves both mirroring and reversal with the monitors mirroring each other so that the clouds appear to be potentially converging, followed by a reversal of the footage, where the clouds appear to retreat from one another in a form of reverse entropy. The piece evokes an inversion of surveillance communication, and in this sense the image of the sky not only depicts weather systems but also alludes to the myriad invisible signals that now traverse our stratosphere. Manglano-Ovalle's focus here on information processing and systems has an important precedence in the recent history of contemporary art. Think of the rise of video art from the beginning of the 1960s or, more obliquely, the interest in systems aesthetics that preoccupied a generation of artists of that era. Critics have drawn links between this interest and new technologies emerging at that time. In 1969, for example, in the 'Software' exhibition at the Jewish Museum in New York, curator Jack Burnham drew parallels between emerging dematerialized art practices such as conceptualism and new information processing systems. The show included the first public exhibition of *hypertext*,[7] presented adjacent to the work of Joseph Kosuth. Edward A. Shanken, in comparing information (content) to computer software, says of Kosuth's work that "one can interpret the artist's propositions as operating like instructions in the mind of the viewer."[8]

Manglano-Ovalle updates this focus on the interaction between systems, technology and communication. In *Wind Shear* he exposes myths surrounding the process of globalization, a development implicitly linked to the vast and now seemingly viral and continuously expanding communications network that engulfs the world. The emergence of this network has led some commentators to discuss notions of a technological sublime. Paul Virilio, for example, sees a relationship between religious and technological cultures, and suggests of the new information technologies that they "bring into effect three traditional characteristics of the divine: ubiquity, instantaneity and immediacy."[9] Similarly, Manglano-Ovalle acknowledges a sense of the overwhelming by making clear that while the new information technologies help to enable globalization, they are also increasingly used as a means of control to keep people apart. In considering *Wind Shear*, it is important to note that it was shot on the US/Mexican border. When the film reverses, the clouds are turned back, echoing the process of border control, which itself can be construed as an attempt to halt an entropic process. The concept of entropy, from the second law of thermodynamics, proposes an irreversible process in a closed system toward equilibrium. While entropy can be arrested or affected by external forces, its irreversible pressure remains. In reminding us that, unless otherwise controlled, weather is not subject to borders and has invisible attributes, *Wind Shear* also alludes to the aggressive march of globalization and the consequence that the production of cultural value is increasingly beholden to the operation of the global market, with an ensuing reduction in the autonomy of communities and nations.

In *Active Military and Reconnaissance Satellites of the United States of America*, 2008, Trevor Paglen provides us with a deeply chilling reminder of the immense apparatus of control facilitated by the new information technologies by depicting the approximately 200 such satellites currently in operation. Henrik Håkansson's looped film projection *Monarch–The Eternal*, 2008, however, suggests a foil to such a meta-system of surveillance. The subject of this work is the Monarch butterfly and the eternal spirit of life and unresolved enigmas suggested by its patterns of migration. Like the weather, the butterfly is unencumbered in crossing borders to complete a migratory cycle that has spanned generations. It remains a mystery how the butterfly passes specific geographic and botanical information to descendants many generations removed. The implication might be that this information is encoded genetically, and in this way Håkansson's work reminds us that genetic research and manipulation at a microcosmic level is radically altering concepts of nature and our understanding of the body. Computers are now being used to decipher the DNA life "code." In turn, such developments in science have resulted in a shift towards thinking of the body as information, information contained at a molecular level. In other words, our understanding of the body is a product of our technological development, and the body is not immune to the process of refiguring all information as code.

On another level, Håkansson's film reveals the community of butterflies as an organism itself, as if the behavior of each butterfly was determined by some larger system. This concept is expanded upon in *La Fabrique*, 2006, by Tania Mouraud, whose installation depicts the industry of the weavers multiplied many times over to suggest a generative system of human connection. However, she establishes a dialectical tension between this idea of connection and the individual by having each weaver stare at the camera and thereby the viewer. We are forced to engage the weavers individually as they continue their rhythmic labour. If our current era is leading us to figure the body as information, countering an earlier mechanistic view of the body, Mouraud's piece also reminds us that manual labour continues to be the mainstay of production for much of the world's population.

References to genetic transfer within larger systems is also an undercurrent in *God is Design*, 2005, by Adel Abdessemed, where what appear to be mutating cells intermittently fuse with potent symbols of Islam, Judaism and Christianity. The film could allude to the theory of memetics, which postulates that units of self-replicating cultural information can be transferred in much the same manner as genetic information. More pertinently, if its political considerations are evident through the suggestion of competition and conflict, it also suggests a much larger quasi-cultural/organic system whereby difference and connectivity are interdependent.

References to belief systems, involving the religious, mystical or supernatural, proliferate within the exhibition in a way that is not absolute in terms of affirmation or critique. Mircea Cantor's *About Airplanes and Angels*, 2008, invokes the supernatural realms of both flying carpets and angels, played off against images of airplanes in formation. In another piece that accompanied the exhibition in the gallery magazine *Switch*, Cantor fuses the idea of DNA coding with the astrological signs of the zodiac. In his photograph *Per Speculum*, 2006, Adrian Paci references the Bible's "First Letter to the Corinthians." In *Mandala: Zone of Zero*, 2003, Kimsooja references Tibetan mandalas in the form of jukebox speakers that emit Tibetan, Islamic and Gregorian chants. Meanwhile, two of Ann Veronica Janssens'

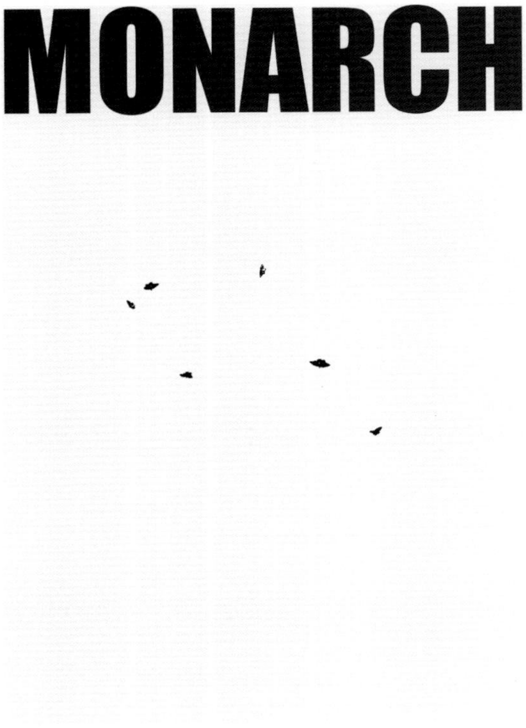

HENRIK HÅKANSSON, entrance poster for *Monarch–The Eternal*, 2008.

pieces are forms of documentation of an eighteenth-century observatory in India that is the largest surviving monument to Vedic mysticism, which draws on the Vedas. Dating back to around 1500 BC, the Vedas are thought to be the oldest sacred texts and are believed to contain the secret laws of the universe. The structure itself, Jantar Mantar in Jaipur, India, is a measuring device and a predictive aid to charting cosmic events, such as eclipses. Janssens' videos act as silent observers, referencing time, space and measurement through poetic record. Nothing specific is elucidated on cosmic belief systems, yet via her camera, she positions herself bodily in relation to their physical and temporal legacy.

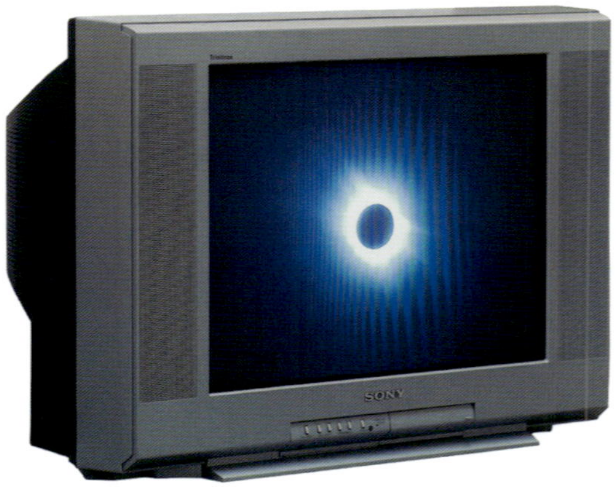

ANN VERONICA JANSSENS, *Side (Natural Version)*, 2006
Single-monitor installation, 3 min., 3 sec., loop, colour, silent.

The body is a central trope considered in different ways by artists in the exhibition. For Hirschhorn the body is the medium that engages the philosophical universe. For Janssens the body is not represented; nevertheless, it also acts as a medium, engaging both the interdependence and the built-in indeterminacy of overlapping systems. With *Side (Natural Version)*, 2006, she silently records a solar eclipse, a transit of her own body, the moon and the sun. The bodily reference is made specific through her dedication of the work to Joseph Plateau, a nineteenth-century Belgian physicist, known for his experiments concerning the persistence of vision after the eyes are closed, who became blind after staring at the sun for twenty-five seconds. In different ways the activated body is the focus of works by Antonia Hirsch and Jed Lind. With *String Theory*, 2003, Hirsch presents an endless loop of a performer skipping, intermittently producing an infinity symbol via the movement of the rope. Lind's photograph *Better to Burn Out No. 1*, 2008, maps the rotation of the earth in relation to the heavens by physically rotating the camera over an extended period of time. The resulting image is less a record of space than it is of time, specifically time in relation to his bodily movements and experience.

In other works the body is suggested more subliminally, as with Abdessemed, or by its absence, as in Paterson's work, a spectre evoked by the self-playing piano. Similarly there is no body sitting at the desk in *Telephone/Time*, but a feature of Cardiff's recorded phone conversation is her reflection on the experience of time as a simultaneous form of embodiment and disembodiment. For Manglano-Ovalle and Paglen, the body is implicitly the object of surveillance, while body parts are illustrated directly in Fred Tomaselli's *Car Bomb*, 2008. This work references the violence and ensuing chaos of political and religious conflict, in an image that presents a vertiginous collision of different representational orders and belief systems, as a version of the Big Bang. In his painting *Roter Platz*, 2007, Franz Ackermann implicitly references the movement of bodies via the nomadic nature of sections of contemporary society. In many other ways the figure of the body haunts the exhibition, at times through its absence or disintegration. Both Tomaselli and Ackermann explore the present moment as chaotic, a mess of information drawn from different and often oppositional systems. In response to this clamour Janssens focuses in, via a seeming quest to isolate the essence of the universal. Yet what appears at first as documentation of an event opens out and resonates in unexpected ways. With *Side (Studio Version)*, 2006, she alters the footage she shot of a solar eclipse, sequencing the sun's corona through various colour transformations. The work recalls modernist abstraction, while remaining a documentation of celestial alignment, presented in an alien light. At one point the colours adopted are those of the Belgian flag, hinting at the cultural contingency of her quest.

Interlocking cultural frameworks are considered in a more expansive way in Tyson's *The Block*, which insinuates the impermanence of culture as much as it explores the impermanence of form. Through his Eurocentric representation of the history of the universe in a series of photographs, culture arrives with the body, which leads to religion, science, the discovery of the new world, imperialism, and modernity as a chain of events. We would expect the next image in the sequence to signify the present moment; what is presented is a flattened image, a diagram representing an abstract system. Form mutates into code, and as a consequence history is dematerialized. The series of images begins and ends with a bronze block. Each representation is further punctuated by an image of molten bronze to suggest the transformation of the block through stages. The fragile order represented by this narrative is belied by the suggestion of impermanence and obliteration. All these multiple physical and cultural references to generation, transformation and eradication are thereby wrapped up in the physical block that Tyson solemnly positions in the centre of the installation. It offers us a particular narrative of the universe encased within the thing-ness and gravity of the block. As matter and form, this block will continue to be subject to entropy and will by implication continue to transform or break down. More optimistically, the form can be read as a building block, a provocation perhaps by Tyson to develop a different and non-deterministic understanding of, and relation to, the universal.

THE IMMEASURABLE INTERMEDIARY

> *If there is no measure, the metaphysicians say, there is no cosmos, and if there is no cosmos there is no state.*[10]
>
> *We will seek another, different social universal. Right away we will have to discover a new object. The money is using itself up, the weapons are at their maximum, the fetishes are dead. I know what the object of science is. But we must find a different object if we want to survive.*[11]

In *Studio Wall Drawing: 29 Dec 2001: Somewhere Near the Edge of the Visible Universe*, 2001, Tyson depicts a star field at the outer limit of the cosmos, beyond which we presume a void. A grouping of stars spells out "Turn back now." The warning is open-ended; nevertheless, it reminds us that we are enclosed within,

KEITH TYSON, *Studio Wall Drawing: 29 Dec 2001: "Somewhere Near the Edge of the Visible Universe . . . "*, 2001
Mixed media on watercolour paper, 157 x 126 cm.

29 DEC 2001 "SOMEWHERE NEAR THE EDGE
OF THE VISIBLE UNIVERSE ....."
TURN
BACK
NOW

and are also part of, the known and radiating universe. Our planet remains our centre. We can imagine and even plot another location in the universe, as Angela Bulloch does in *Night Sky: Mars from Venus. 12*, 2008 or as The Otolith Group do in *Otolith*, 2003 a semi-fictional tale of a future extraterrestrial human community, which can only learn about the earth, their forbears and their history via the media and archive. Both works remind us that our understanding of the eternal, of history, of the universe and our place within it, is entirely mediated.

The concept of a mediated universe is a constant in the work of Gabriel Orozco. He has often considered the construct of the universe in relation to the internal logic and aesthetics of a game. In the work *Ping Pond Table*, 1998 for example, he produces a four-sided Ping-Pong table in the form of a quatrefoil, a form that has a potent history of signification, beginning with a representation of a flower with four petals. Each wing of the table connects to create a square in the centre, which contains a lily pond. The pond is a generative system, a universe unto itself perhaps, but a system that can be interrupted. Across this pond two sets of players potentially play across one another. One imagines that a given set of players would focus on their game, but would be potentially distracted by the other players' ball, or interrupted by the pond via their ball landing on and rippling its surface. The implication here is that any given system is ultimately open to interruption. This idea is more implicit in Orozco's work *Black Kites Perspective (front horizontal)* from 1997. The engraving on the skull follows the form of a chessboard. Computers now can predict every possible move within the game of chess, making it an entirely decipherable system. However, no computer can yet predict an interruption to a game – a phone call perhaps or a knock at the door. Orozco allows for the coherency of a system, but also points to an infinite multiplicity of systems that have the potential to disturb and even transform each other. Measurability and certainty decline as the network of systems expands.

As with many of the artists in 'Universal Code,' Orozco positions the universe not as a system but as an intermediary. Using this idea, connectivity is explored by the artists in the exhibition as an interacting set of relationships between our evolving understanding of the cosmos, the production of scientific and biological knowledge, systems aesthetics, cultural belief systems, information technologies, and global power relations. 'Universal Code' considers the response of the artists to these relationships in the aftermath of globalization, reflecting the complexity of the world we currently inhabit. It charts the renewed interest of artists in concepts of the cosmological and immeasurable in an era where there is no longer a fixed scale that measures value, thereby producing a consequent loss of cultural and scientific certainty. Ultimately the response of the artists is poetic, positioning the universe as a void full of potential but also as a field riddled with elision and enigma.

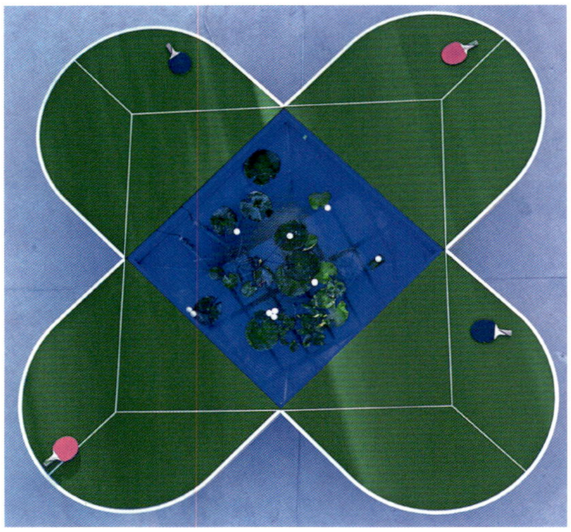

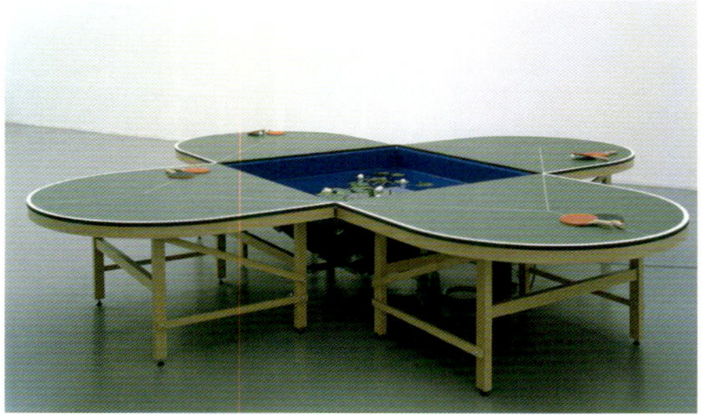

GABRIEL OROZCO, *Ping-Pond Table*, 1998
Modified ping pong tables, water lilies, water, mixed media, 30 x 167 3/4 x 167 3/4 in., 76.2 x 426.1 x 426.1 cm.

NOTES

1. Michel Serres, *Genesis*, trans. Geneviève James and James Nielson (Ann Arbor: University of Michigan Press, 1995), 107.

2. Michel Serres, *Genesis*, trans. Geneviève James and James Nielson (Ann Arbor: University of Michigan Press, 1995), 97.

3. Fernand Hallyn, *The Poetic Structure Of The World: Copernicus and Kepler* (New York: Zone Books, 1993), 236.

4. Alison Sky, interview with Robert Smithson, "Entropy Made Visible," 1973, originally published in Nancy Holt, ed. *The Writings of Robert Smithson* (New York, New York University Press, 1979). Sourced from www.robertsmithson.com.

5. Alison Sky, interview with Robert Smithson, "Entropy Made Visible," 1973, originally published in Nancy Holt, ed. *The Writings of Robert Smithson* (New York, New York University Press, 1979). Sourced from www.robertsmithson.com.

6. Michel Serres, *Genesis*, trans. Geneviève James and James Nielson (Ann Arbor: University of Michigan Press 1995), 103.

7. Hypertext is an interactive computer-based system that utilizes a text-linking structure, the most familiar example of which is Hypertext Markup Language (HTML), used for the World Wide Web to encode character- and text-formatting information.

8. Edward A. Shanken, "Art in the Information Age: Technology and Conceptual Art," in *Leonardo* 35, no. 3 (August 2002), 435.

9. John Armitage, interview with Paul Virilio, "From Modernism to Hypermodernism and Beyond" in *Virilio Live: Selected Interviews*, ed. John Armitage (California: Sage Publications, 2001), 36.

10. Michael Hardt and Antonio Negri, *Empire* (Cambridge, Mass.: Harvard University Press, 2000), 355.

11. Michel Serres, *Genesis*, trans. Geneviève James and James Nielson (Ann Arbor: University of Michigan Press 1995), 91.

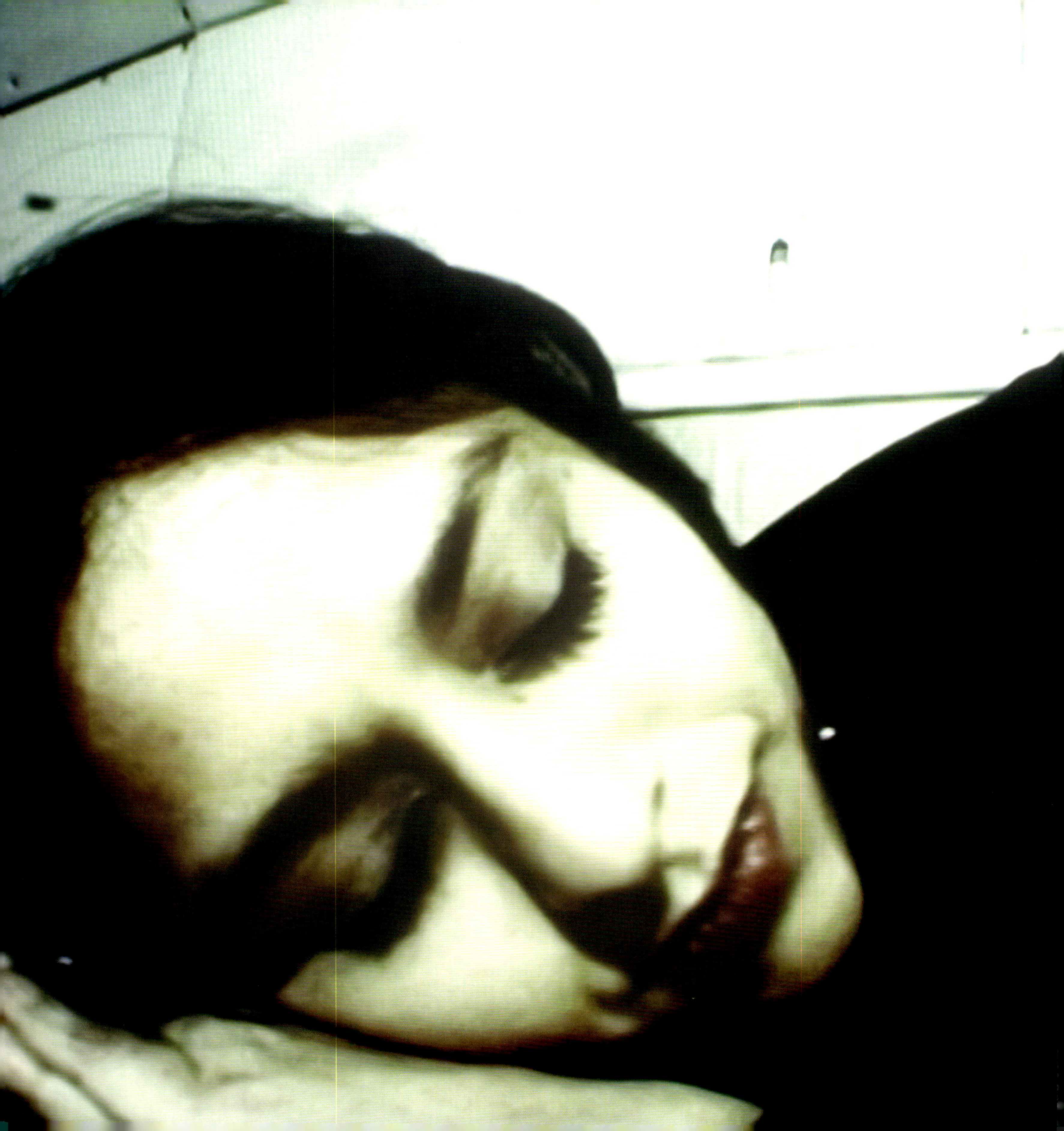

ECSTATIC UNIVERSE: BATAILLE'S NEGATIVE ECOLOGY

JANINE MARCHESSAULT

UNIVERSAL CODE

"I am the emptiness of caskets and the absence of myself in the whole universe."
Georges Bataille

IS THIS WORLD, MADE OF PEOPLE WITH NOTHING IN COMMON, BASED ON some invisible universal communication? Where might such communication take place – at the level of the molecular and the cosmological, in the fabric and patterns of deep time? Can the ecstasy of laughter and the "tears of eros" provide the affective space of total communication – our bodies as the ground for a utopia of communication, for a universal code that is drenched in time? More than any artist-philosopher of the twentieth century, Georges Bataille was driven by such questions, all of which are raised in the profoundly beautiful and probing exhibition 'Universal Code.'

Jean Baudrillard tells us that capitalism severs the universe from the code, transforming the "universal code" into a matrix of consumption, a form that is transparent (codes that we live by, codes that determine, delimit and communicate our social standing), reducing human beings to consumers and social relations to objects:

> *"It is a totalitarian code, for no one escapes it; escaping it in a private sense cannot prevent us from participating every day in its collective development. Not believing in it still means believing sufficiently in other people's belief in it to adopt a skeptical stance. Even actions intended as resistance to it must be defined in terms of a society that conforms to it."*[1]

For Baudrillard, this system is driven by a fundamental fear of death, which is effaced and codified as the "non-living" in the science of biology: "Death undertakes to abolish death, and for this purpose erects death."[2] Death becomes a static object of study, a category and dividing line in the positivist enterprise.

Indeed, Bataille's last book, *The Tears of Eros*, 1961, written in ill health (he died a year after its publication) and anxiety because his eldest daughter had been arrested for her political activities on behalf of Algeria, seeks to understand this fundamental relation between sex and death as precisely a challenge to scientific codifications: "The ambiguity of this human life is really that of mad laughter and of sobbing tears. It comes from the difficulty of harmonizing reason's calculations with these tears . . . With this horrible laugh . . ."[3] His book brings together an historical archive of images – drawings, engravings, paintings, and photographs that engage in one way or another with the meeting of eroticism (orgasm/*le petit*

THE OTOLITH GROUP, *Otolith*, 2003, still/detail
Video installation, 22 min., 20 sec., colour, sound.

mort) and death across several cultural enclaves. It is a book (censored when it first appeared) which is aimed at the "opening up of consciousness" of this relation through the visual traces of its pulsions. The book is aimed at eliciting an emotional response through its juxtaposition of images. Bataille entreats us to know the world through death in life (opened by eros), to consider new epistemologies that are founded on *ends* rather than on means, and where seemingly opposite affectations (e.g., repulsion/attraction) become indistinguishable.

LAUGHTER

With Bataille, everything begins with the ecstatic tears of laughter.

While undertaking research at the British Museum in 1920, a young Bataille was preparing to meet a philosopher – Henri Bergson, the first philosopher he had ever met in fact. Not wanting to appear ignorant, and for the sake of expediency, Bataille selected one of the shortest books ever written by the philosopher, a small book called *Laughter*. As Bataille's biographer describes the experience, the book, "like the philosopher himself, was a disappointment." Yet the topic was a profound "discovery."[4]

For Bataille, laughter would be a central interest, reoccurring in his diverse body of writing across forty years. Indeed, Bataillean laughter bridges the short period of devout Catholicism in his life to his most profound criticisms of Christianity and his theory of religion via Nietzsche. His reflections on eroticism, sacrifice, ecstasy, and his notion of expenditure resonate deeply in an age where the phenomenologies of planetary and economic catastrophe appear as unspeakable totality.

Laughter has many faces: "When all is said and done, I have more than one face. I don't know which is laughing at which."[5] Laughter is fundamentally a matter of different kinds of communication. Bataille distinguishes between two experiences of laughter. The first is a communication "linking up two beings." A mother tickling her child, for example, elicits ordinary unmediated laughter. The second is a "communication with death." This is what Bataille terms "infinite laughter," a communication with "our beyond," with what he has called the "impossible."[6]

During the First World War, Bataille's syphilic (paralyzed and blind) father died alone in Rheims, a city under siege by the German Army. He had been abandoned by his family who were forced to leave him. His death haunts Bataille's writings. Death is infinite continuity, which for Bataille is a negative reality that we touch for a "lightening" moment through activities linked to sacrifice, laughter or forms of eroticism and ecstasy. These represent a "field of possibilities" tied to assertions of negative thought. This unknowable or undefined reality is not linked to anything that is determined by the supernatural or by God, but rather belongs to "ecstasy in which the divine form melts like sugar and water."[7]

With laughter in particular, there is shared knowledge and a common object which is known from the outside and depends on the individual's history, and culture. Joking involves the merging of two distinct systems, a kind of montage of circumstances that melt into each other – Bataille frequently uses liquid metaphors to conjure the fluid and dynamic boundaries of reality. With the joke, an isolated system comes into contact with another circumstance. I will laugh, Bataille argues, on two conditions: 1) that it occurs suddenly and 2) that our response is not inhibited. The relation between the isolated system and the circumstance it encounters creates a new reality (an event) and Bataille's description of laughter is temporal – it occurs in the moment, and in the encounter between two elements. Take the example of the mother tickling her child. The mother's touch "ruptures" the child's boundaries, the child's sense of inside and outside. The tickling is a "current" connecting them. Laughter is in many ways reducible to the child's "laugh of recognition," as if "what controlled the child, falls into its field."[8]

When we look at each other and laugh, we "slip into one another" in an erotic energy. Communication through language is structured through our concern for things, for objects outside ourselves. Yet laughter creates intimate communication, it is a rupturing of the exterior to reveal our shared interiority. Laughter spreads like a wave throughout a room. It "reverberates," it is contagious, it can be amplified from person to person – laughter sometimes provokes laughter, it becomes its own object. "Full communication resembles flames – the electrical discharge of lightning." Laughter unfolds in a field of movement, not in a container of emotions. It is a burst of energy that erupts suddenly and is shared. It is a transition, a contact and a "slippage" that involves the heterogeneous: I bring another into myself as "*mine* but at the same time *other*."[9] It precisely in this mix that opposites become indistinguishable:

> *"What the hearty laugh screens us from, what fetches up the bawdy jest, is the identity that exists between the utmost in pleasure and the utmost in pain: the identity between being and non-being, between the knowledge which brings one before this dazzling realization and definitive, concluding darkness. To be sure, it is not impossible that this truth itself evokes a final laugh; but our laughter here is absolute, going far beyond scorning ridicule of something which may be repugnant, but disgust for which digs deep under our skin."* [10]

We are not compenetrated by each other (or rarely so) but by the comic object itself, which is in movement. In the fusion ("the waters of two torrents mixing"), the comic elements are exhausted. We cannot remain in a convulsive state of pleasure/pain, outside of ourselves, reduced to an impersonal sensation through this intimacy, endlessly.[11] Communication through laughter can connect us to the continuity of death only momentarily. We cannot keep laughing at the same joke. In this way, this experience is completely temporal and historical.

UNEMPLOYED NEGATIVITY

Bataille's philosophical project is epistemological, empirical and ethical. That is, he practices along with fellow traveller Roger Caillois "Sacred Sociology," also called heterology. Under the rubric of the Collège de Sociologie (College of Sociology), this is:

> *the study not only of religious institutions but of the entire communifying movement of society. Hence it regards power and the army, among other things, as its rightful object, and it contemplates all human activities – sciences, arts, and technology – insofar as they have a communifying value, in the active sense of the world, that is to say, insofar as they are creators of unity.*[12]

The College drew inspiration from the work of Emile Durkheim and his brilliant and devoted nephew, the anthropologist Marcel Mauss, whose lectures Bataille and other Surrealists attended in the early 1930s. For Bataille, it was particularly Mauss's essay on the gift, *Essai sur le don, forme archaïque de l'échange*, 1925 that would be significant. Both Durkheim and Mauss saw a situation in which the social bonds of community and the sacred had become increasingly eroded. Following Durkheim, the founders of the College saw the need to reactivate a new secular form of the sacred within the modern world that worked against the structures of bourgeois individualism. It was the function of sacred sociology to both create and study forms of the sacred. This might indeed have proved an impossible task given that the sacred is a purely negative space. But the College was established to create new kinds of knowledge, heterologies that were grounded in the non-teleological, the unknowable and formless – precisely what does not get into the annals of traditional forms of disciplinary knowledge.

Bataille had invited the preeminent Hegel scholar Alexandre Kojève, who had been his teacher at the École des Hautes Études, to deliver lectures at the College. In 1937, a few days after one of Kojève's lectures, Bataille wrote a letter to his former teacher in which he takes issue with Hegel's notion (and Kojève's interpretation) of "the post-historical world." For Hegel the post-historical represents a time when the process of dialectics, the process of work and negation, finally culminates in a harmonious world that has reached completion. Bataille is concerned about the function of negation in a world in which history has reached an end. Negation, he writes, "disappears or remains in a state of 'unemployed negativity.'"

Hegel's romantic "dark night," which both attracted and repelled Bataille, was always a contained negativity. And this is Bataille's fundamental criticism of the dialectical system – it is not open to life (or death): "I imagine that my life – or, better yet, its aborting, the open wound that is my life – constitutes all by itself the refutation of Hegel's closed system."[13] Negativity has turned away from action because there is nothing left to do or rather "nothing is done any more."[14] The "unemployed negativity" metamorphoses into a work of art that can only be fundamentally "elusory," that "answers no specific situation" and as such is negativity objectified. In this way, there is something to do – the unemployed negativity becomes a "recognized negativity" engaged in "a process of awareness." Bataille describes a project that uses science as the middle term, bringing:

> *"into play representations extremely charged with emotive value (such as physical destruction or erotic obscenity, an object of laughter, of physical excitation, of fear, and of tears). But at the same time these representations intoxicate him, he strips off the straightjacket that has kept them from contemplation and he sets them objectively within the eruption of time that nothing changes. He understands then, that it is his good, not his bad, luck that brought him into a world where there was nothing left to do, and he offers what has become now, despite himself, to be recognized by others . . . And what he has "to do" is to satisfy the portion of existence that is freed from doing: It is all about using free time."*[15]

Such a project, Giorgio Agamben suggests, is a kind of "epilogue" in which "human negativity is preserved as 'remnant.'"[16] Such an undertaking can be seen to have culminated in Bataille's final book *The Tears of Eros*, an historical anthropological/archaeological collection of representational "remnants" of the relation between death and eroticism across the ages. It must be differentiated from Theodor W. Adorno's hermetically sealed negative aesthetics, which resists instrumental reason through a refusal to communicate. Bataille suggests representations that are "communifying," that elicit responses, force us into a "process of recognition." Such an endeavour is all about "free time" but also freeing time.[17]

Though Bataille still relies upon a dialectical system (on the fundamental oppositions of life and death, eroticism and horror, luxury and waste, etc.), it is through the practice of negation that he proposes a practice of philosophy, philosophy as a practice, theory as a processing that is incomplete, unfinished, unknowable, and open. It is not surprising that Bataille would have rejected Bergson's philosophy of time since, in it, the temporal fluidity of duration is seemingly inaccessible and undifferentiated. With Bergson, consciousness is condemned to be alienated, separated, from the temporal realm of duration. Bataille writes in a manner that is not anathema to Bergsonianism: "We are discontinuous beings, individuals who perish in isolation in the midst of an incomprehensible adventure, but we yearn for our lost continuity."[18] But unlike Bergson's notion of "intuition," which enables some form of access to the realm of duration, Bataille reclaims and redeems this lost continuity through the fusion with another body or object in extreme moments where the singular self dissolves into unity, like "one wave lost in a multitude of waves."[19] And laughter like eroticism are pathways into the unknowledge, the sacred or the beyond.

Bataille is closer to Gaston Bachelard, one of Bergson's staunchest critics for whom "coupure,"[20] like the Bataillean "rupture," opens duration to the world through violent eruptions, discontinuities, interiority, moments of experience, and fantasy. Bataille's thought does not so much open thinking and experience out to encompass a phenomenal terrestrial world (though this – the body, words, representations, the earth – is central to his methodology) as is the case with Bachelard and other world cosmologists such as Alfred Whitehead. Instead, he points us to the celestial universe as a whole and as a negativity. He is interested not just in humankind but in the planet's life – in the connection between stars and cells and galaxies and societies, through a negativity that has no direction, that is in movement, in the unending process of materializing the limits of human existence.

The Collège de Sociologie was short lived (1937–39) and as the Second World War was clearly inevitable by the end of the decade, its members dispersed. Agamben notes that the Nietzschean "Declaration" of 1938 issued by the College "betrayed its impotence" when it reduced the international crisis[21] to bourgeois individualism and "man's devirilization . . . so deprived of destiny, that they find themselves absolutely defenseless when faced with the possibility of death, who having no profound reasons to fight, find[ing] themselves inevitably cowards in the face of battle . . . conscious sheep resigned to the slaughterhouse."[22] Indeed, Walter Benjamin, who had attended lectures at the Collège de Sociologie,[23] warned the group that "the metaphysical and political incommunicability" – 'acephalean atheology' – risked playing into the hands of a 'prefascist aestheticism.'"[24] Such concerns had been directed at the Surrealists by both Benjamin and Brecht earlier in the 1930s.

Bataille's writings on fascism, Hitler and democracy during the 1930s drew inspiration from his dedication to a particular notion of community. Indeed, Denis Hollier has located Bataille's project to consolidate a new "community of men" under the rubric "Sorcerer's Apprentice," in terms of traditions of German Romanticism, which furnish us with some of the first models of an avant-garde practice.[25] Central to these models is the creation of the community, which is prioritized over individual authorial signatory "which allowed the resurgence of the anonymous enunciation, belonging to great periods of community – 'The art of writing in common,' said Novalis, 'is a curious symptom that it is the presentiment of great progress in literature'."[26] As Hollier explores, Bataille's own brand of community with the College was based on a "virile unity" between men: "Women can put their clothes back on: The threat of war would do the job as well as they. Virile Unity is not the uniting of man and woman. It is man's unity confronted with woman [or war]. Virility, like life, is whatever does not let itself be cut apart, whatever resists partition."[27] Hollier adds a caveat: "Jean-Luc Nancy is right: What is at stake in such groups is not the fusion of finitudes in an us. We have to think of it, rather, as a mechanism of erasure, a machine for desubjectified, impersonal enunciation."[28] Indeed, Nancy's work on community draws inspiration from Bataille. What is original in Bataille's concept is the erasure of community, a notion of community founded on *disappearance* and *the impossible*. He continues: "Bataille is without doubt the one who experiences, most acutely, the modern experience of community as neither a work to be produced, nor a lost communion, but rather as space itself, and the spacing of the experience of the outside, of the outside-of-self."[29] It is the unemployed community or, as Nancy puts it, the "inoperative [*désoeuvré*] community" which Bataille proposes that perhaps offers new models for thinking about the world as a plural community consisting of those who have nothing in common.

CELESTIAL HOLE

It is to Bataille's notion of community that we return in the planetary age defined by the limits of energy and expenditure. His two volumes of *The Accursed Share* – read now as deeply insightful, prophetic even – present a challenge to the political economy of capitalist society through a concept of "General Economy." Here society and nature, the interior and exterior of expenditure, are not differentiated. His reflection upon consumption concentrates on its non-utilitarian aspects, on what he terms "unproductive expenditure." Like unemployed negativity, this unproductive experience of consumption focuses on the exorbitant, the wasteful, the luxurious, on excessive emotions and useless activities such as "mourning, war, cults, the construction of sumptuary monuments, games, spectacles, arts, perverse sexual activity." These activities have no meaning, or end, outside of themselves. Any form of consumption that is a means to an end (for example, to stave off hunger) is not a form of expenditure. Rather, expenditure is precisely that which is unproductive.

Bataille's system is based on his knowledge of the history of expenditure (he was, after all, a medievalist librarian) and a notion of general economy which he derives from the idea of potlatch, forms of gift-giving tied to social life, which Marcel Mauss had studied in pre-capitalist tribes. The phenomenon of potlatch is articulated through complex social rituals in which expenditure is privileged over acquisition. Such relational forms are completely foreign to the linear rationality of production and restriction that defines capitalist modes of production. Bataille's conceptualization of general economy then begins by proposing an *integrated* view of energy expenditure. Taking the car industry as an example, he tells us that it is possible to change a tire without considering "the whole, of which the tyre . . . is an integral part." Yet he goes on:

> *Between the production of automobiles and the general movement of the economy, the interdependence is rather clear, but the economy taken as a whole is usually studied as if it were a matter of an isolatable system of operation. Production and consumption are linked together, but,*

considered jointly, it does not seem difficult to study them as one might study an elementary operation relatively independent of that which is not . . . In other words, isn't there a need to study the system of human production and consumption within a much larger framework?[30]

Bataille calls for a transformation of the ethics upon which capitalist economies have been built. He proposes a shift from a restrictive economy (for example, an economy based on scarcity) to a general economy that does not simply praise economic growth without considering the very surface of the earth. He proposes a cosmic ecology in which processes of energy expenditure (economies) and the very nature upon which these processes depend are considered together. Finally:

An immense industrial network cannot be managed in the same way that one changes a tyre . . . It expresses a circuit of cosmic energy on which it depends, which it cannot limit, and whose laws it cannot ignore without consequences. Woe to those who, to the very end, insist on regulating the movement that exceeds them with the narrow mind of the mechanic who changes a tyre.[31]

In reading *The Accursed Share*, published in 1949, one is struck by a seeming shift from Bataille's other writings on eroticism and sacrifice to a focus on the material life of human existence, and on the radical proposal for rethinking political economy in ecological terms. Yet there is a consistent project here: the "unemployed negativity," the unproductive consumption, like the capacity for slippage and excess, enables thinking outside of oneself to disclose the world in relation to a universe beyond. Love for Bataille is always exorbitant, filled with risk, chance and loss, and defined by the universal: "What I love in the person I love . . . is the universal aspect of that person."[32] This love of the universal, the universal as love, temporalizes (historicizes) the "Code" in an aleatory and endless process of communication. It is without doubt this unthinkable world that fuels an irreducible holism in Bataille's philosophy, a non-dialectical world of relationalities and interdependencies, whose difficult totality demands of us an ethics of the planetary — an ecology which is expansive, radiant and open.

NOTES

1. Jean Baudrillard, *The System of Objects*, trans. James Benedict (New York: Verso, 1996) 212–13.
2. Jean Baudrillard, *Symbolic Exchange and Death* (London: Sage Publications, 1993), 152.
3. Georges Bataille, *The Tears of Eros* (1961), trans. Peter Connor (San Francisco: City Lights, 1989), 20.
4. Michel Surya, *Georges Bataille: An Intellectual Biography*, trans. Krzysztof Fijalkowski and Michael Richardson (London: Verso, 2002), 36.
5. Georges Bataille, "Hegelian Concepts," in *Bataille: A Critical Reader*, eds. Fred Botting and Scott Wilson (Oxford, U.K.: Blackwell Publishers, 1998), 94.

6. Georges Bataille, "Laughter," in *Bataille: A Critical Reader*, eds. Fred Botting and Scott Wilson (Oxford, U.K.: Blackwell Publishers, 1998), 59. See also Georges Bataille, *Guilty*, trans. Bruce Boone (San Francisco: The Lapis Press, 1988), 139–43.

7. *Ibid*., 59.

8. *Ibid*., 60.

9. *Ibid*., 62.

10. Georges Bataille, "Madame Edwarda," in *The Bataille Reader*, eds. Fred Botting and Scott Wilson (Oxford, U.K.: Blackwell Publishers, 1997), 225. See also Georges Bataille, *My Mother, Madame Edwarda, The Dead Man*, trans. Austryn Wainhouse (London: Marion Boyars, 1989), 137–59.

11. *The Bataille Reader,* 61.

12. Georges Bataille and Roger Caillois, "Sacred Sociology and the Relationships between 'Society,' 'Organism,' and 'Being,' Saturday, November 20, 1937," in *The College of Sociology* (1937–39), ed. Dennis Hollier and trans. Betsy Wing (Minneapolis: University of Minnesota Press, 1988), 74.

13. *Ibid*., "Hegelian Concepts, Alexandre Kojève, Saturday, December 4, 1937," 90.

14. *Ibid*., 92.

15. *Ibid*., 91–92.

16. Giorgio Agamben, *The Open: Man and Animal*, trans. Kevin Attell (California: Stanford University Press, 2004), 7.

17. *Bataille: A Critical Reader*, 92.

18. Georges Bataille, *The Unfinished System of Nonknowledge*, ed. Stuart Kendall and trans. Michelle Kendall and Stuart Kendall (Minneapolis: University of Minnesota Press, 2001), 15.

19. "Sacred Sociology" in *The College of Sociology* (1937–39), 75.

20. Cf. Gaston Bachelard, *L'intuition de l'instant* (1932) (Paris: Gonthier, 1971).

21. The international crisis was referring to the Munich Agreement of 1938 — six months later the German troops entered Czechoslovakia.

22. Georges Bataille, "Declaration of the College of Sociology on the International Crisis," in *The College of Sociology* (1937–39), 45.

23. As is by now well known, Walter Benjamin had entrusted the folder containing *Passagenwerk* to Bataille, who hid it at the Bibliothèque nationale.

24. Pierre Klossowski, *College of Sociology*; "Entre Marx et Fourier," *Le Monde*, 31 May 1969.

25. Cf. Philippe Lacoue-Labarthe and Jean-Luc Nancy, *The Literary Absolute: The Theory of Literature in German Romanticism*, trans. Philip Barnard and Cheryl Lester (Albany, N.Y.: SUNY Press, 1988).

26. In the foreword to The College of Sociology, Hollier criticizes Bataille's virile aesthetics (xvii).

27. *Ibid*.

28. *Ibid*., xxi

29. Jean-Luc Nancy, *The Inoperative Community*, ed. Peter Connor and trans. Peter Connor, Lisa Garbus, Michael Holland, and Simona Sawhney (Minneapolis, MN: University of Minnesota Press, c. 1991).

30. "The Meaning of General Economy: The Dependence of the Economy on the Circulation of Energy on the Earth," *The Bataille Reader,* , 182–83. See also Georges Bataille, The *Accursed Share, Volume 1: Consumption*, trans. Robert Hurley (New York: Zone Books, 1988).

31. *The Bataille Reader*, 186–87.

32. "Love," *The Bataille Reader*, 96.

ADEL ABDESSEMED
FRANZ ACKERMANN
ANGELA BULLOCH
MIRCEA CANTOR
JANET CARDIFF &
GEORGE BURES MILLER
CERITH WYN EVANS
HENRIK HÅKANSSON
ANTONIA HIRSCH
THOMAS HIRSCHHORN
ANN VERONICA JANSSENS
KIMSOOJA
JED LIND
IÑIGO MANGLANO-OVALLE
JOSIAH MCELHENY
TANIA MOURAUD
GABRIEL OROZCO
THE OTOLITH GROUP
ADRIAN PACI
TREVOR PAGLEN
KATIE PATERSON
FRED TOMASELLI
KEITH TYSON

THE EXHIBITION

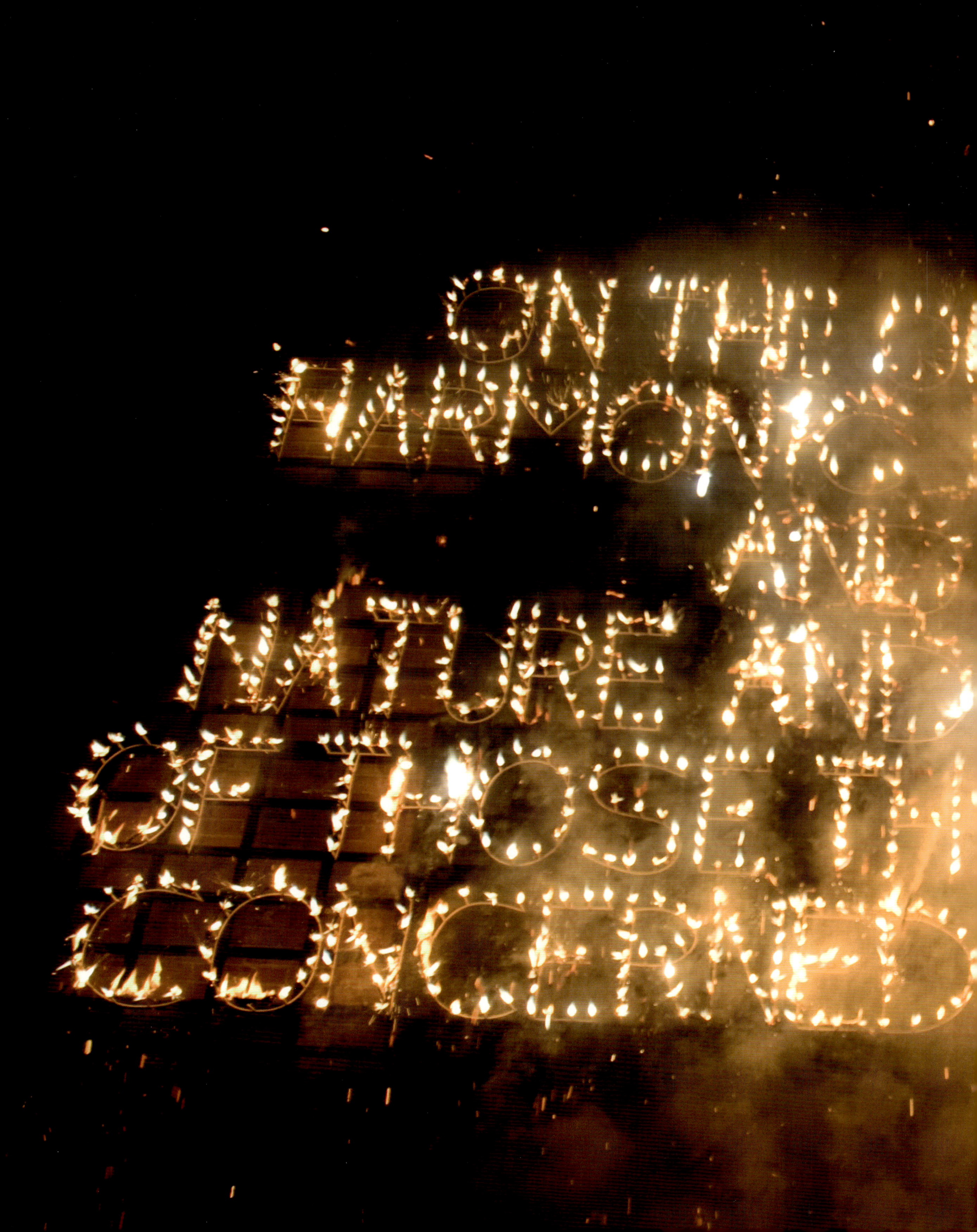
ON THE
HARMONY
NATURE
OF THOSE
CONCERNED

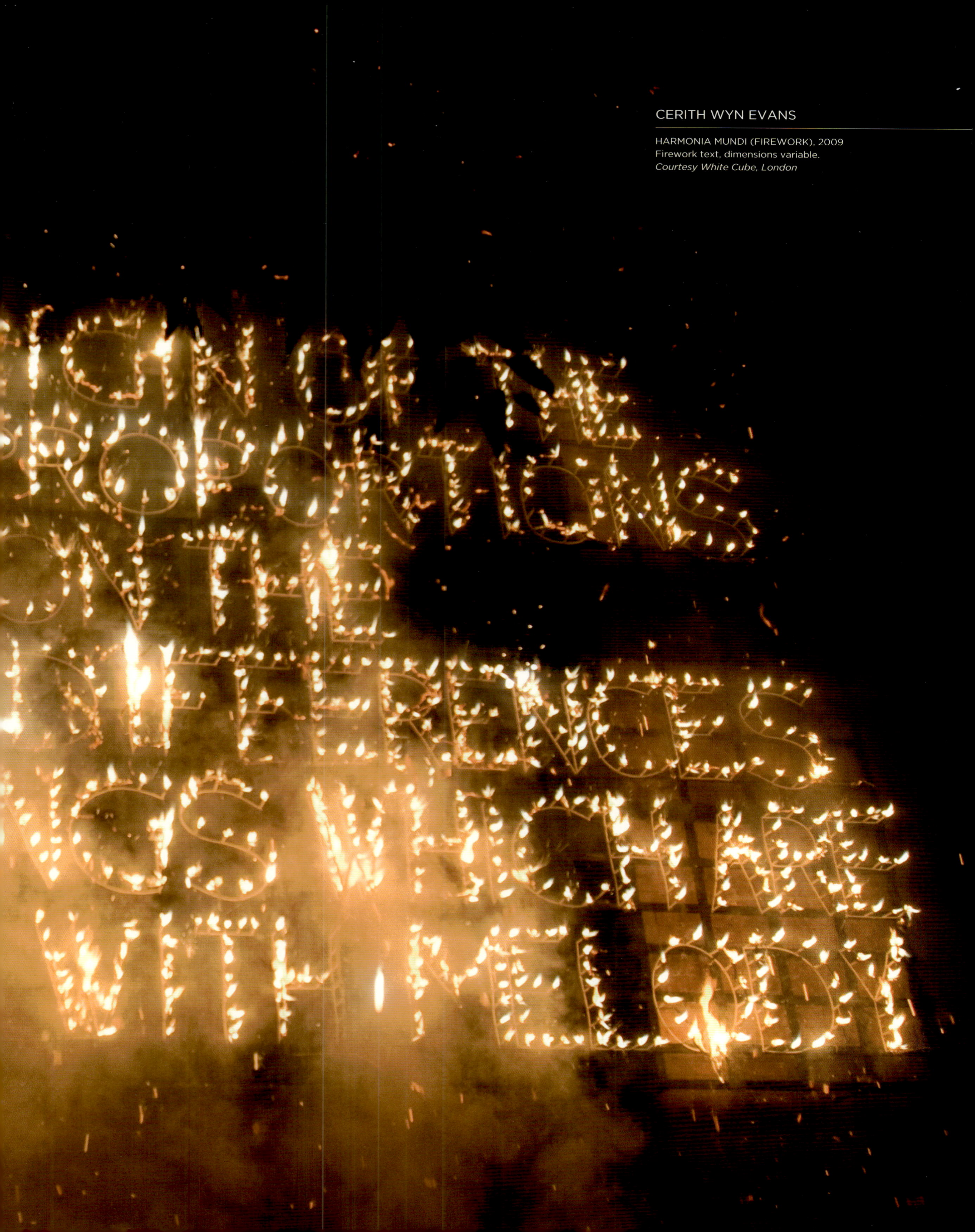

CERITH WYN EVANS

HARMONIA MUNDI (FIREWORK), 2009
Firework text, dimensions variable.
Courtesy White Cube, London

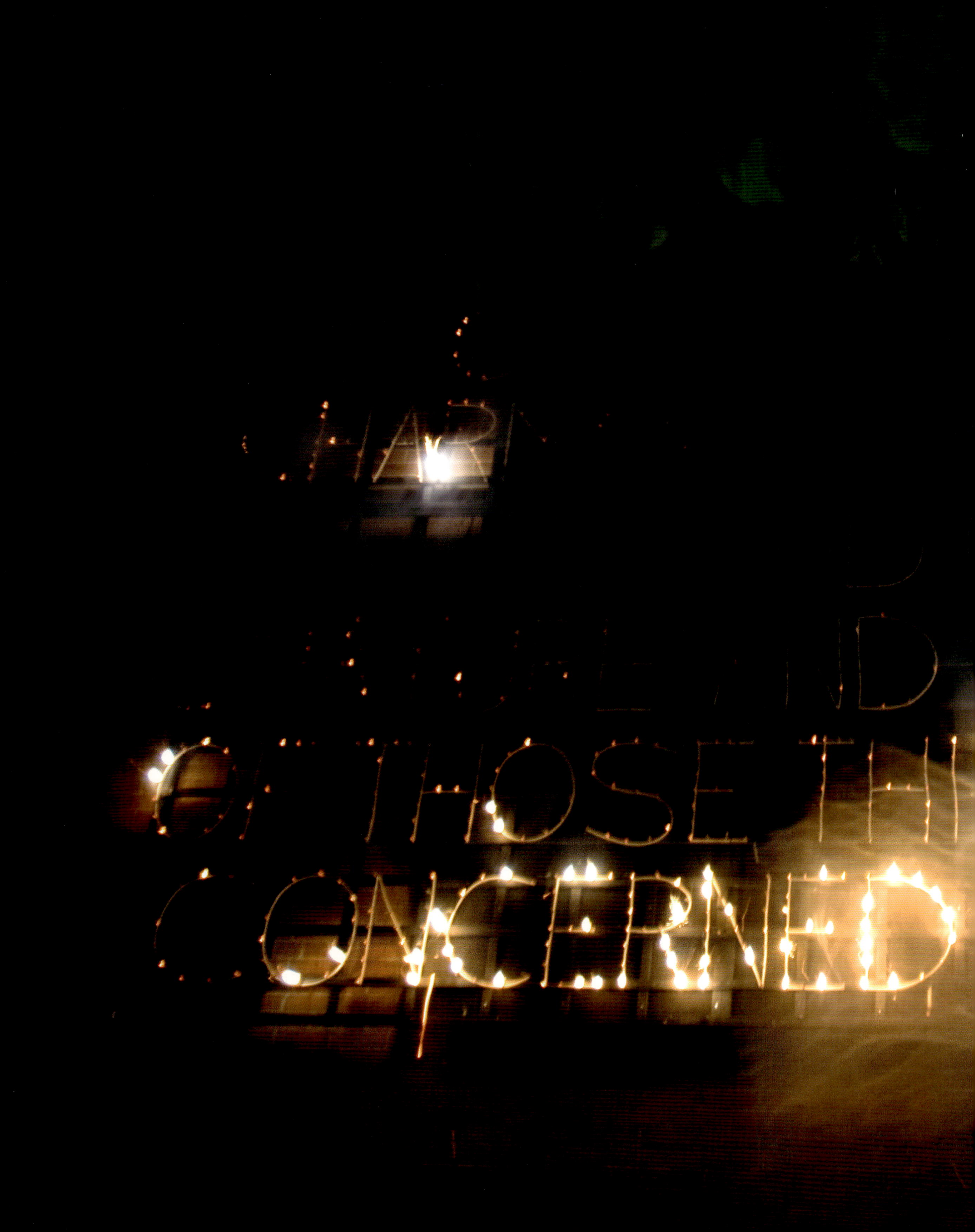
CONCERNED

HARMONIA MUNDI (FIREWORK), 2009

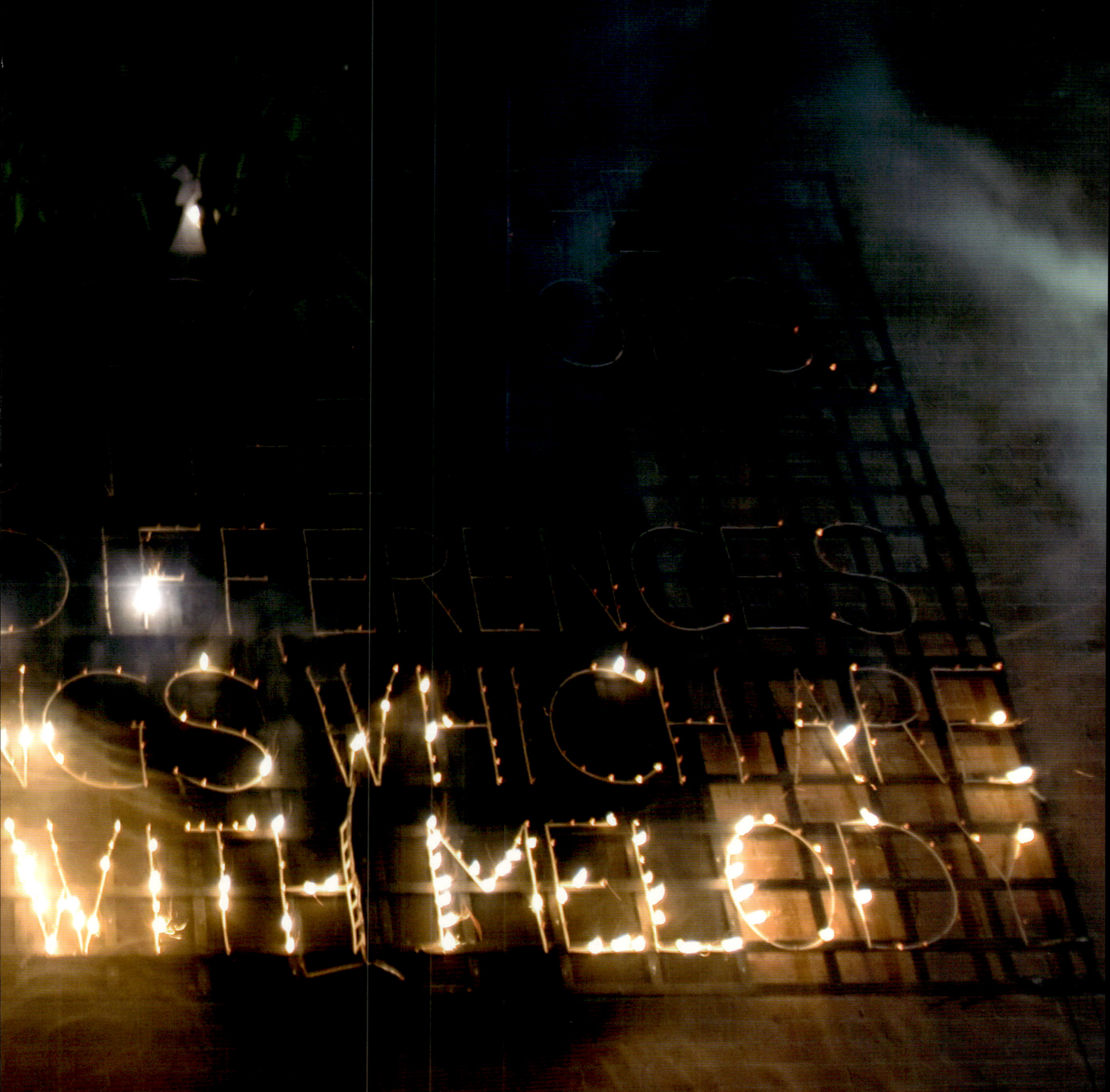

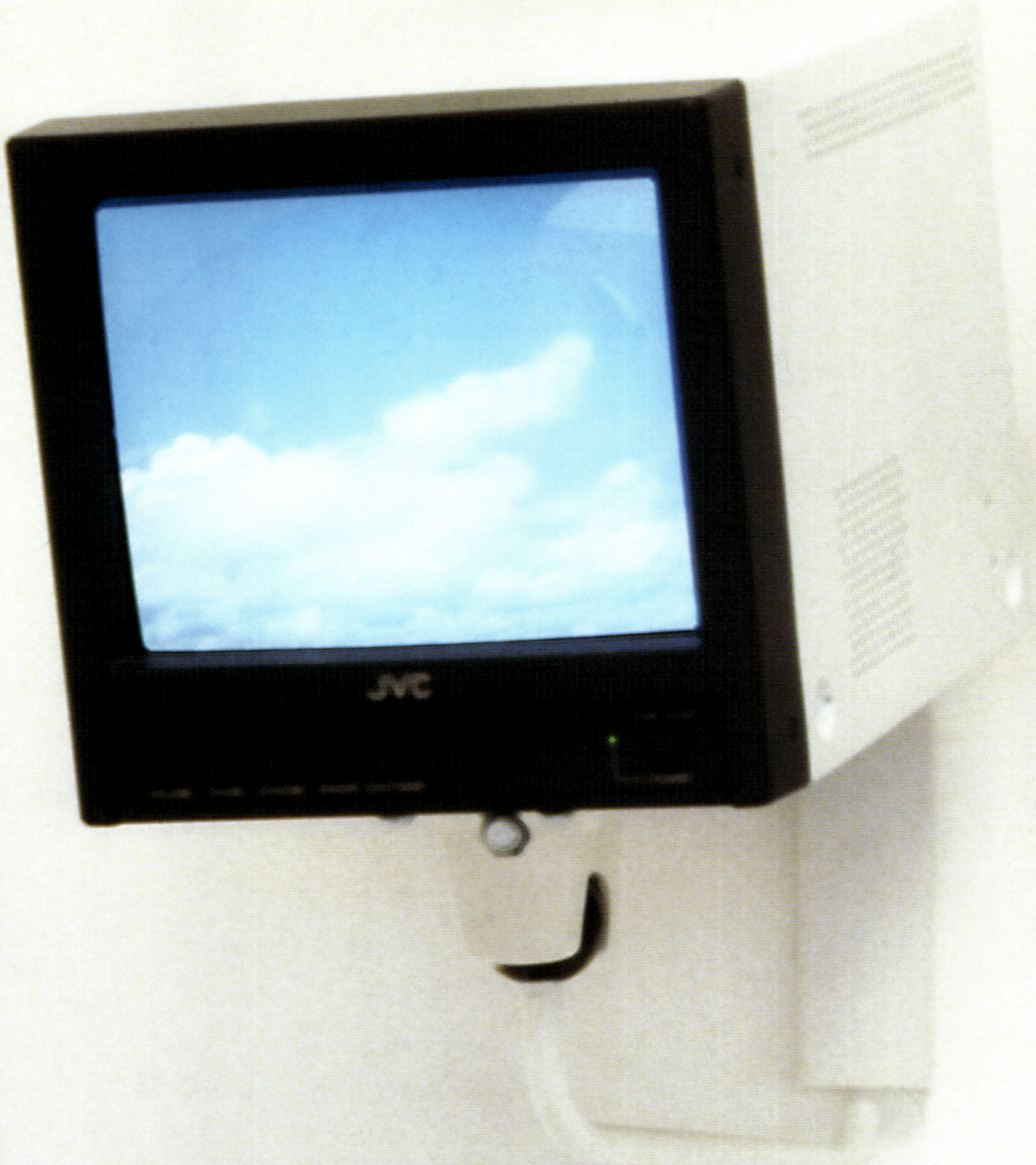

IÑIGO MANGLANO-OVALLE

WIND SHEAR, 1997
Two-channel video monitor installation,
loop, colour, silent.
Collection Thomas H. Bjarnason

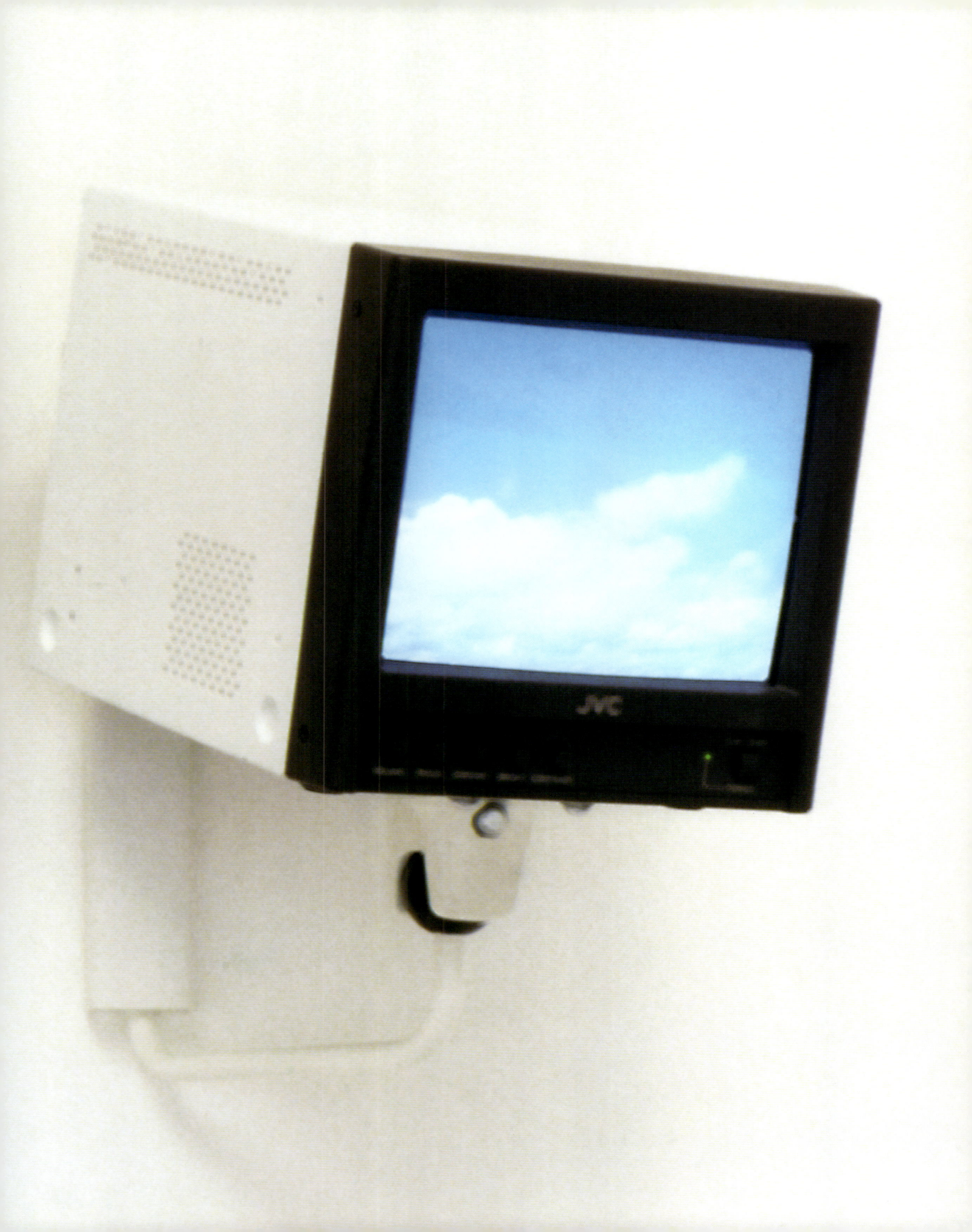

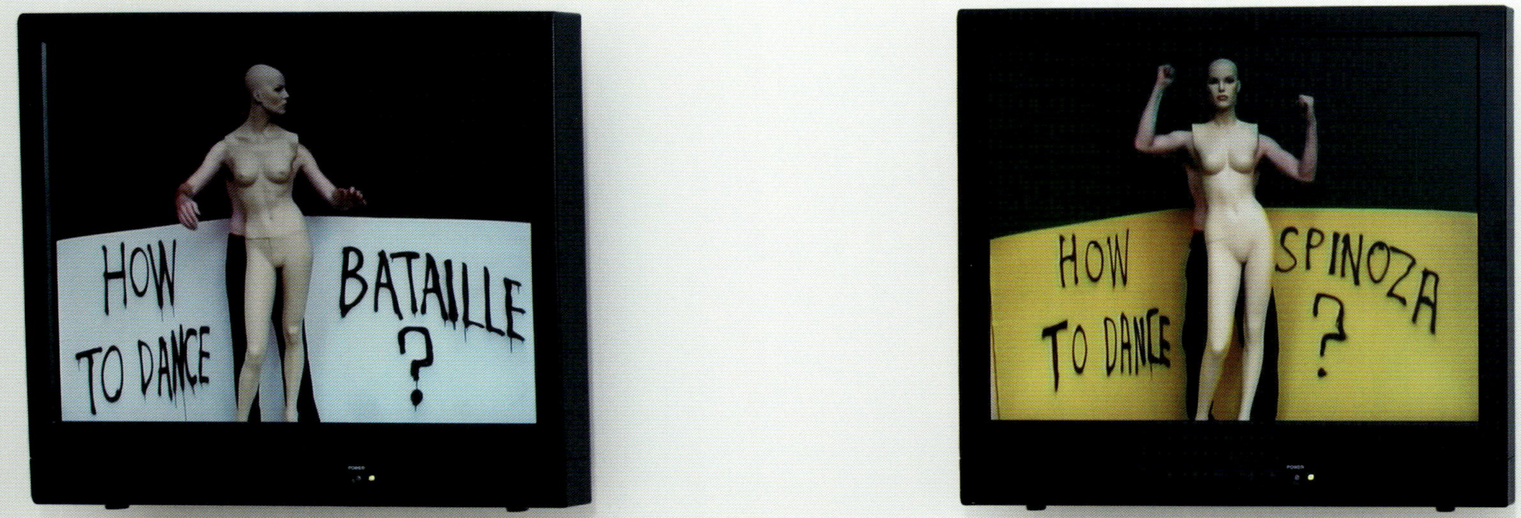

THOMAS HIRSCHHORN

DANCING PHILOSOPHY
(HOW TO DANCE DELEUZE?),
(HOW TO DANCE GRAMSCI?),
(HOW TO DANCE BATAILLE?),
(HOW TO DANCE SPINOZA?), 2007
Video installation, 5 min., 25 sec.; 3 min., 37 sec.;
4 min., 5 sec.; 4 min., 23 sec; loop, colour, silent.
Courtesy Galerie Chantal Crousel, Paris

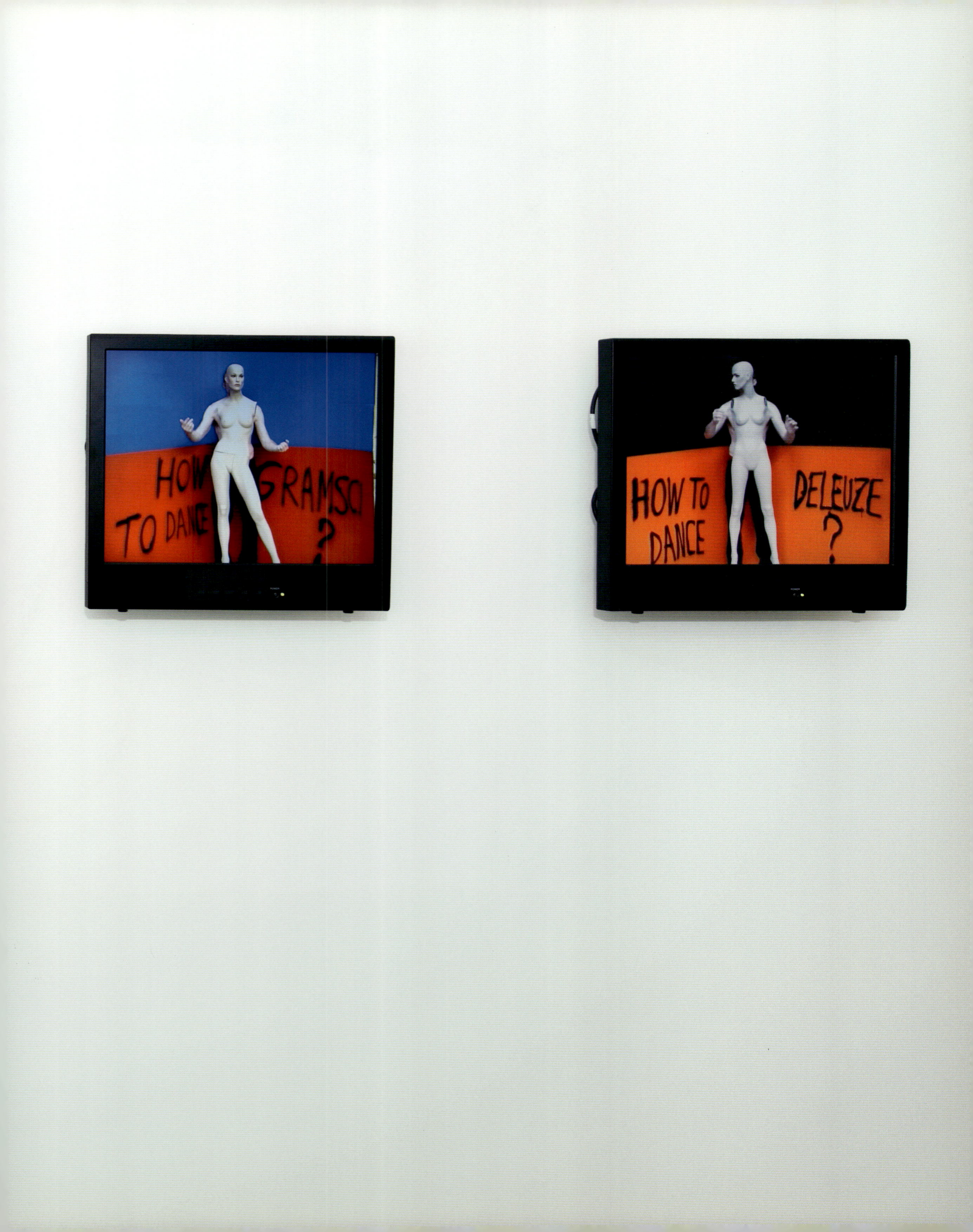
HOW
TO DANCE
GRAMSCI
?
HOW TO
DANCE
DELEUZE
?

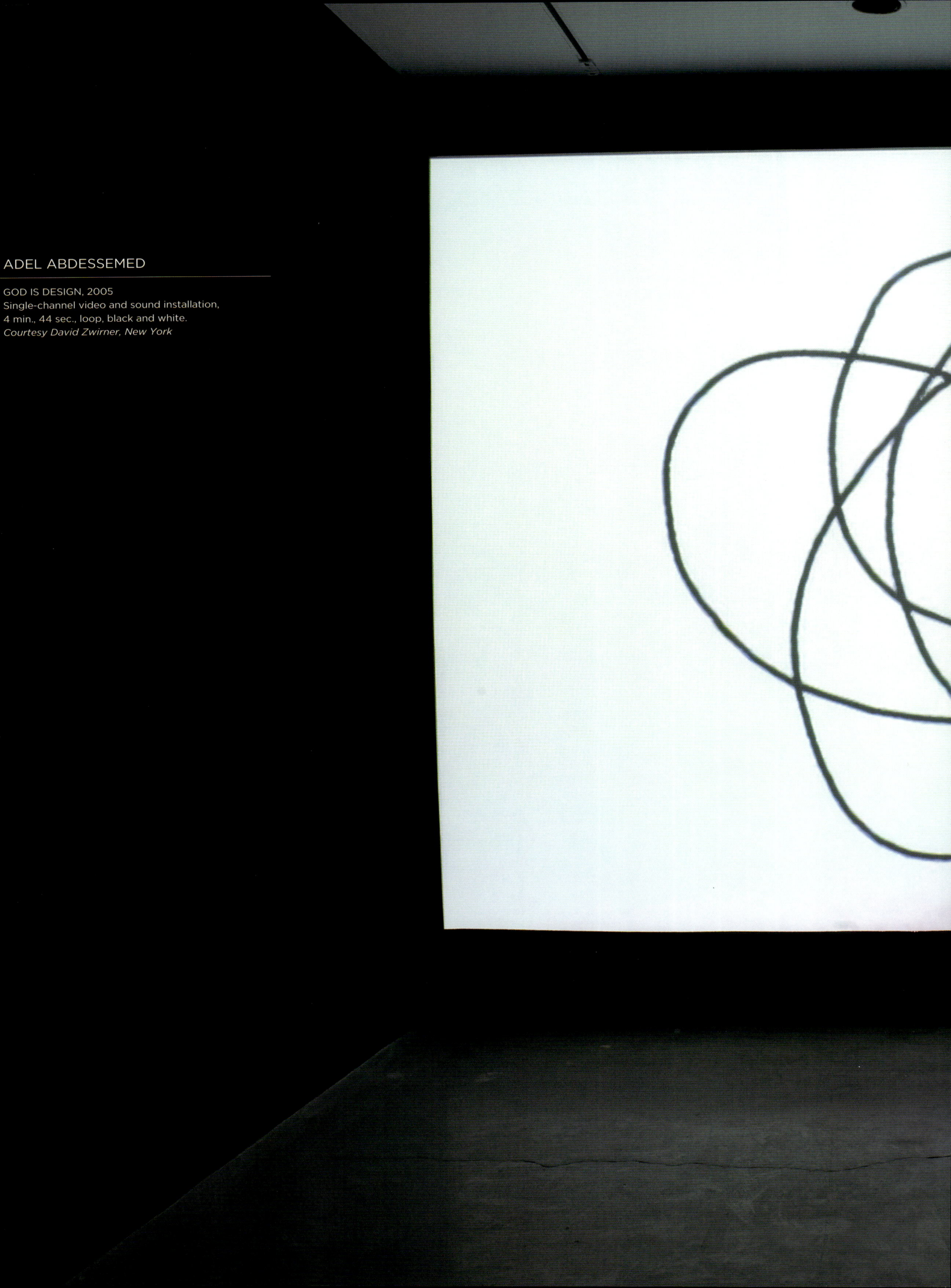

ADEL ABDESSEMED

GOD IS DESIGN, 2005
Single-channel video and sound installation,
4 min., 44 sec., loop, black and white.
Courtesy David Zwirner, New York

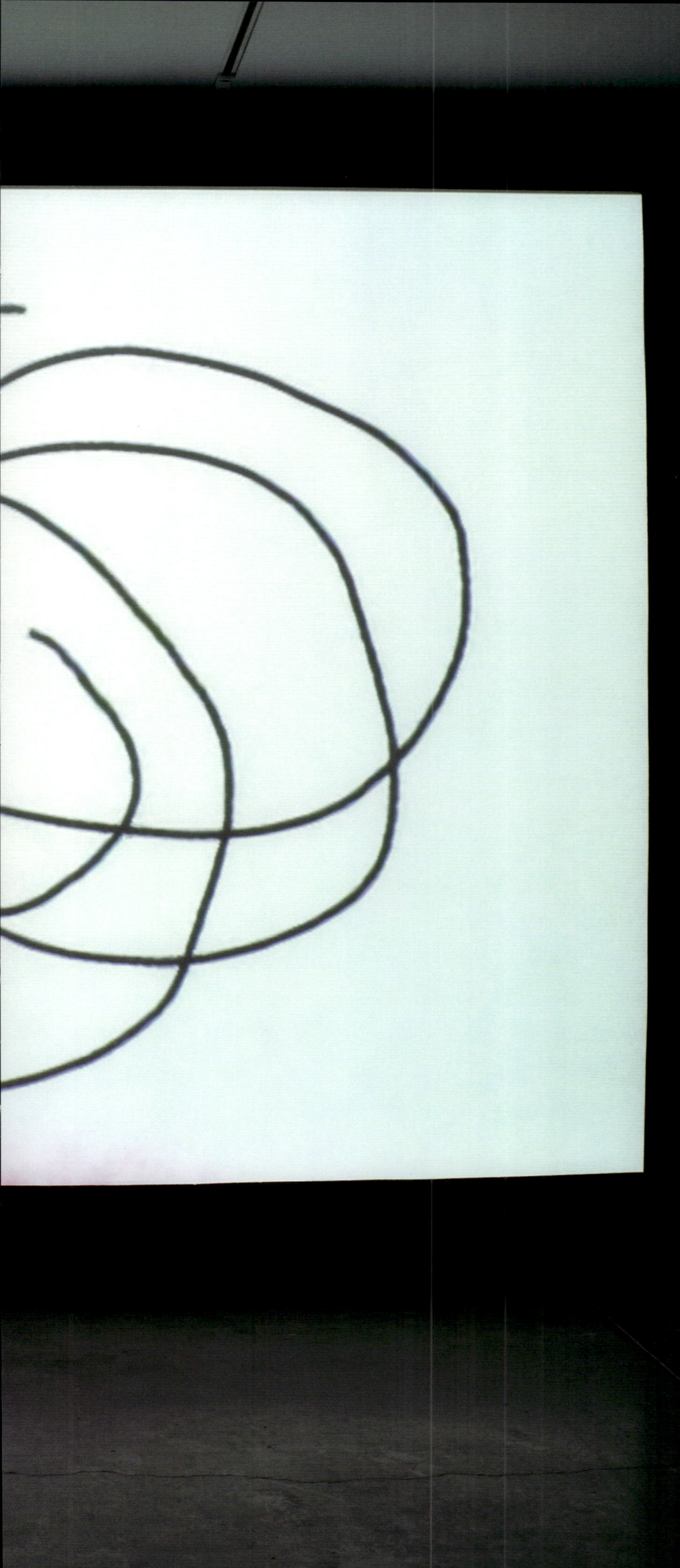

ADEL ABDESSEMED

GOD IS DESIGN, 2005

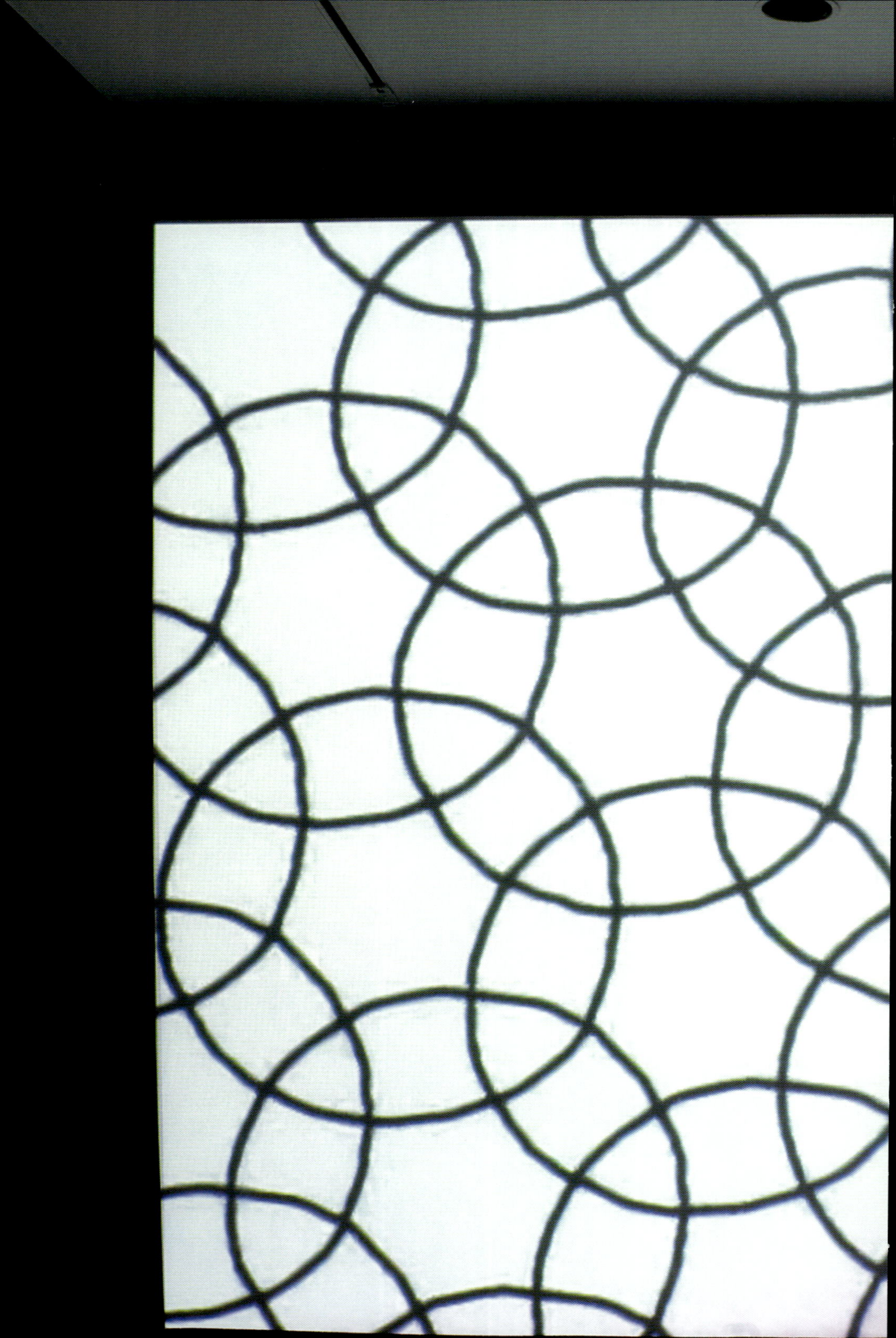

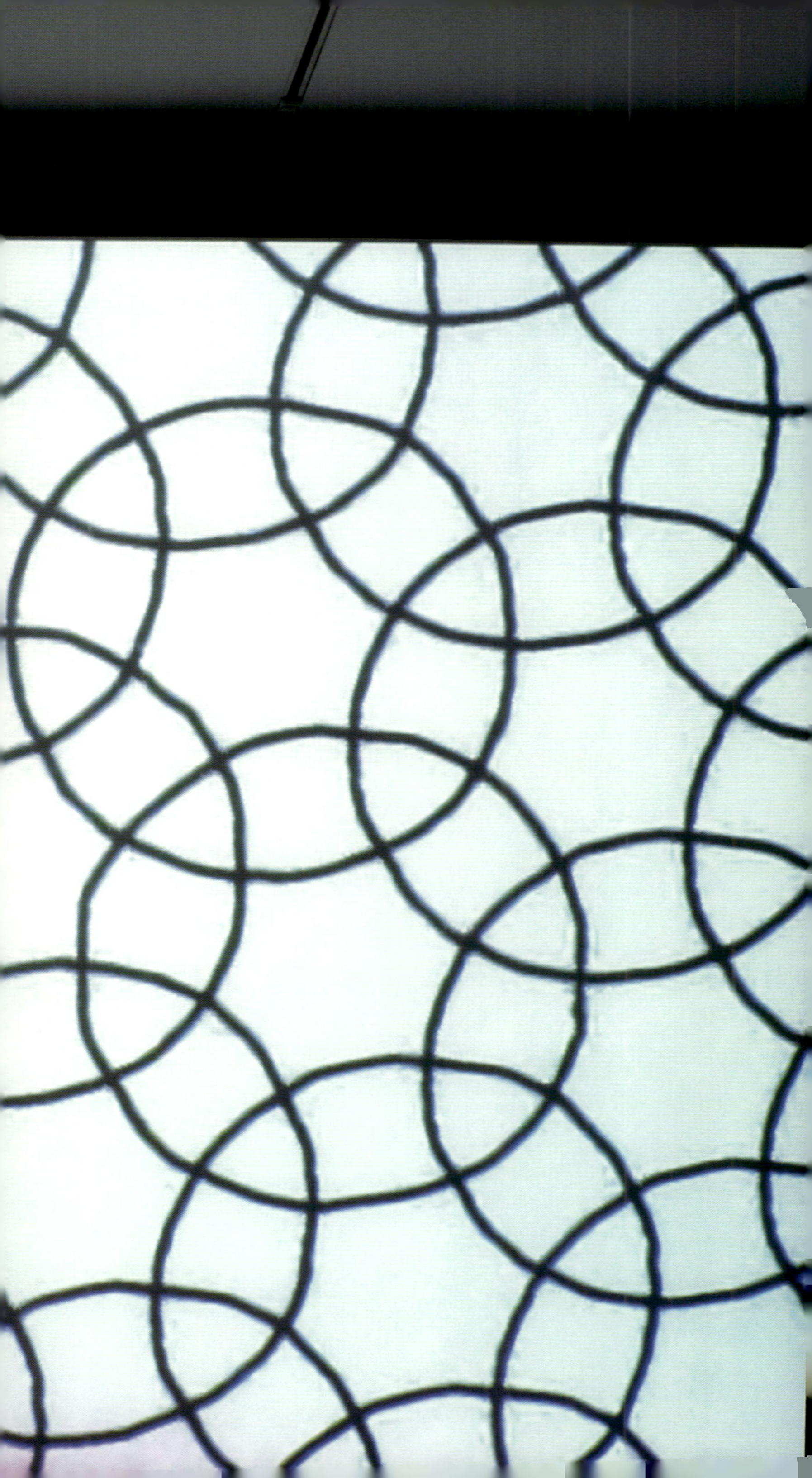

ADEL ABDESSEMED

GOD IS DESIGN, 2005

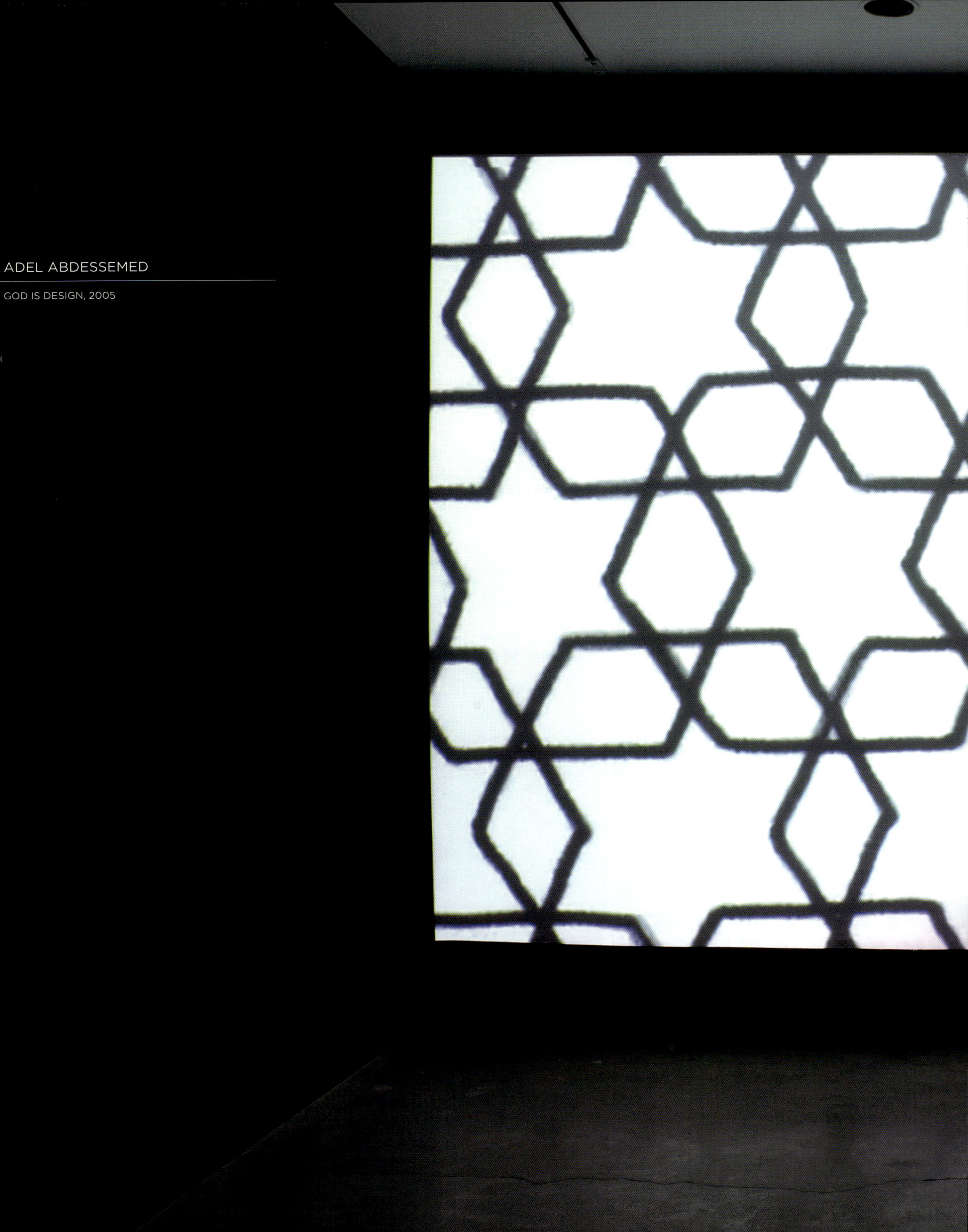

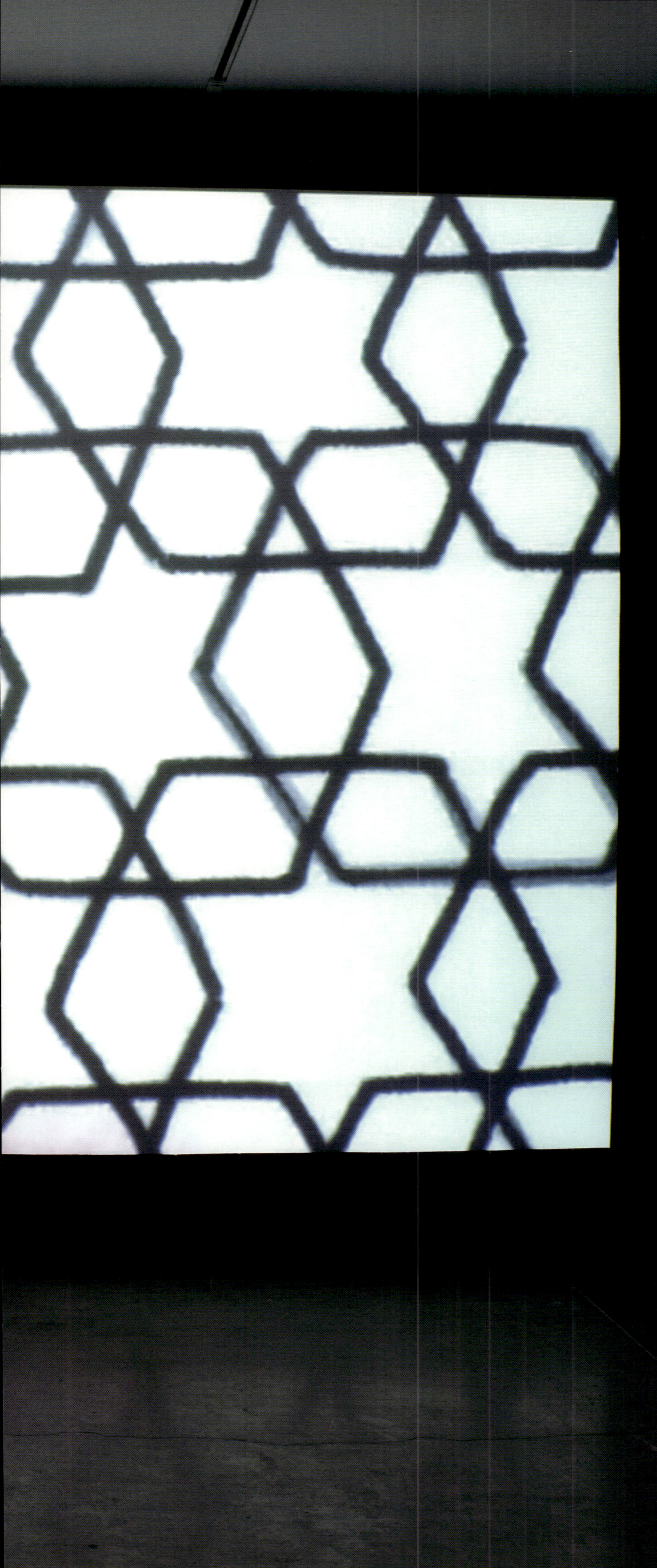

ADEL ABDESSEMED

GOD IS DESIGN, 2005

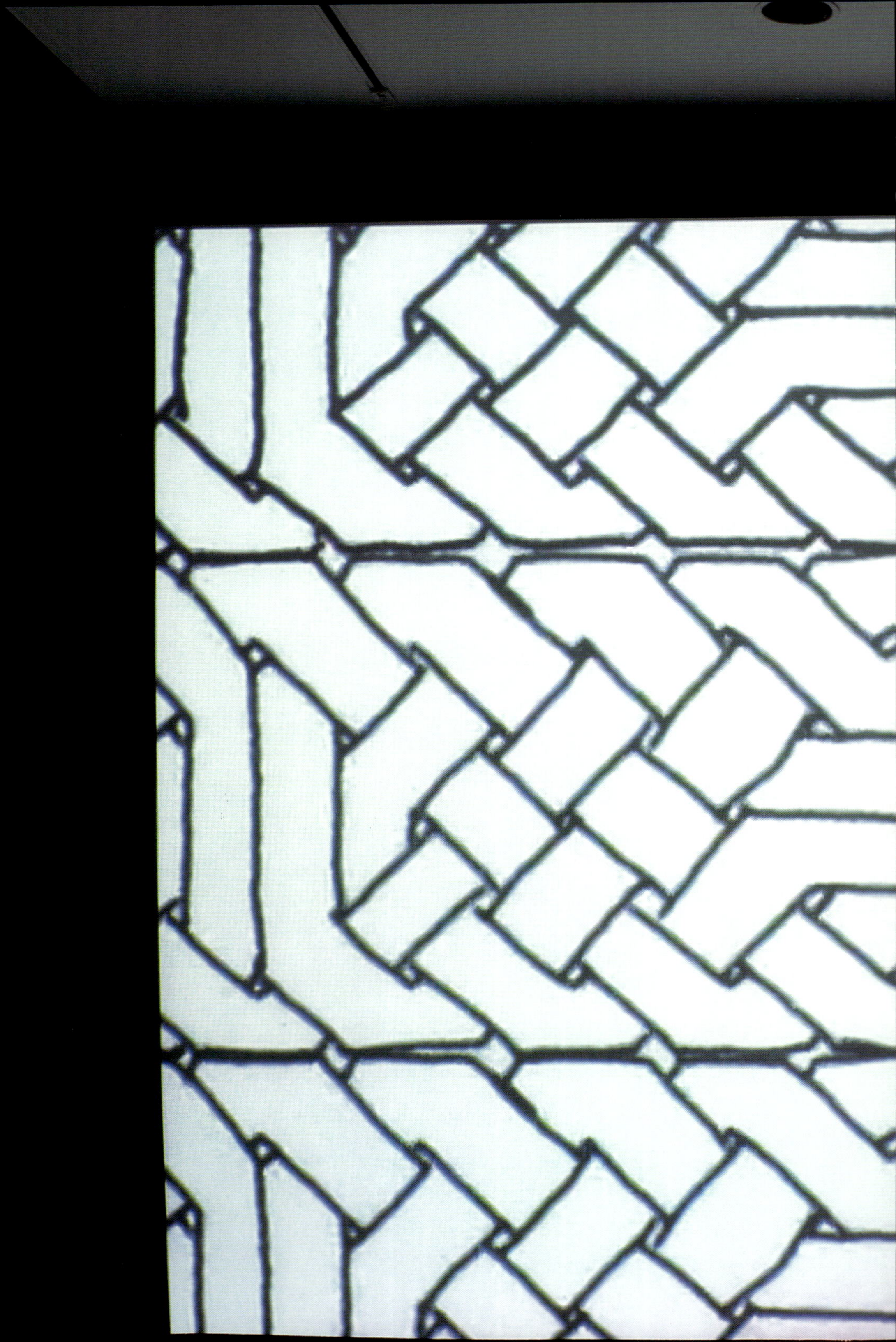

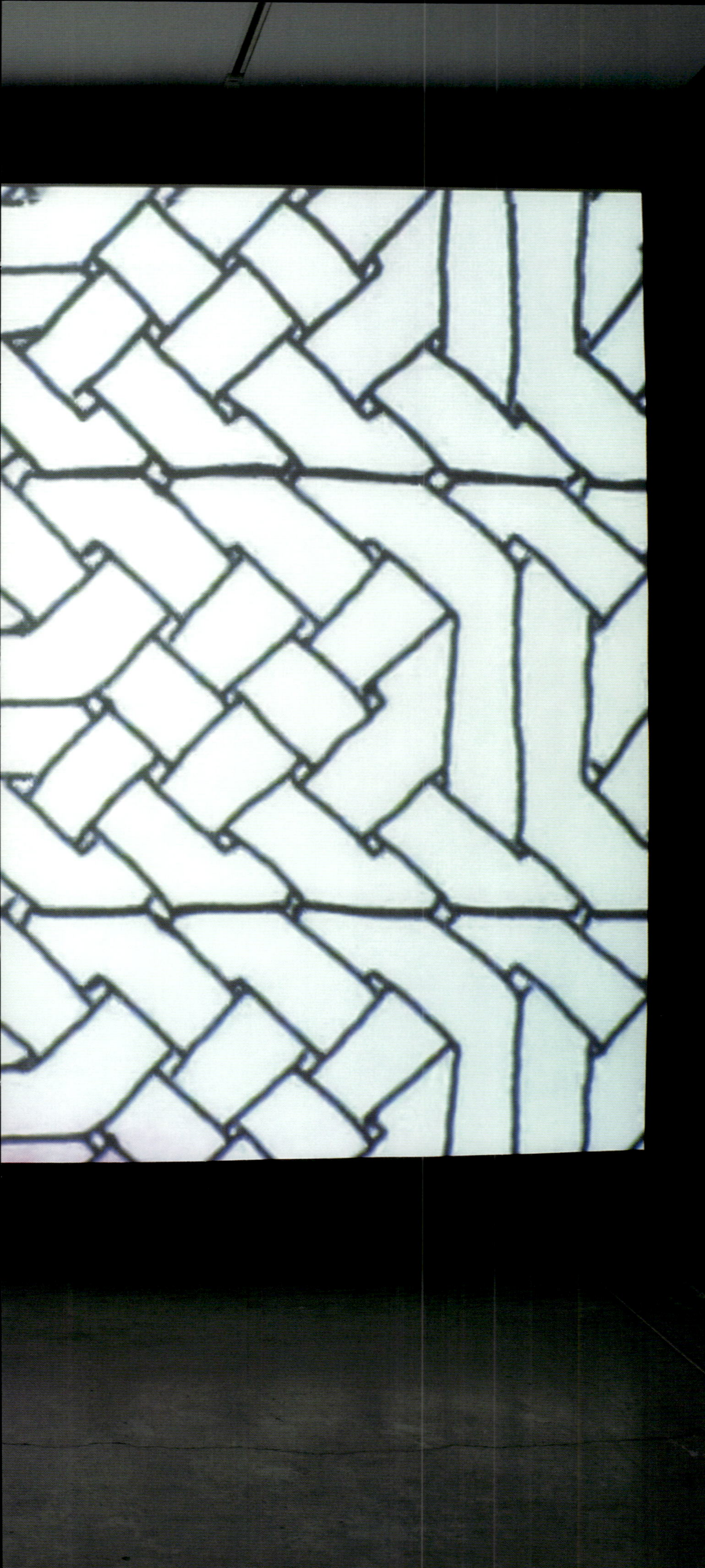

THE OTOLITH GROUP

OTOLITH, 2003, still
Video installation, 22 min., 20 sec., colour, sound.
Courtesy of the artists

THE HUMAN NEUROVESTIBULAR SYSTEM IS SEVERELY COMPROMISED BY sustained periods in microgravity. The otoliths – the part of the inner ear that senses tilt – are especially affected. Faced with lack of stimulation of the otoilths in the ear's semicircular canal, the saccule and the utricle begin to expand. In the course of the 5 missions to the International Space Station carried out between 2009 and 2070, the deformation of the otoliths during foetal development prevented newly born children from learning to manage Earth's gravity. Prolonged duration in microgravity has led to a bifurcation in hominization. Microgravity has adapted a new human species. We are better suited to space than gravic humans could ever be; but we are unable to survive conditions on earth. Earth is out of bounds for us now; it remains a planet accessible only through media.

NOTES FROM REPORT ON HUMAN ADAPTATION TO PERMANENT AGRAVIC HABITATION
AS PRESENTED TO THE INTERNATIONAL SPACE STATION INSTITUTE OF EXOANTHROPOLOGY.

DR. USHA ADEBARAN-SAGAR.
24 · 10 · 2103

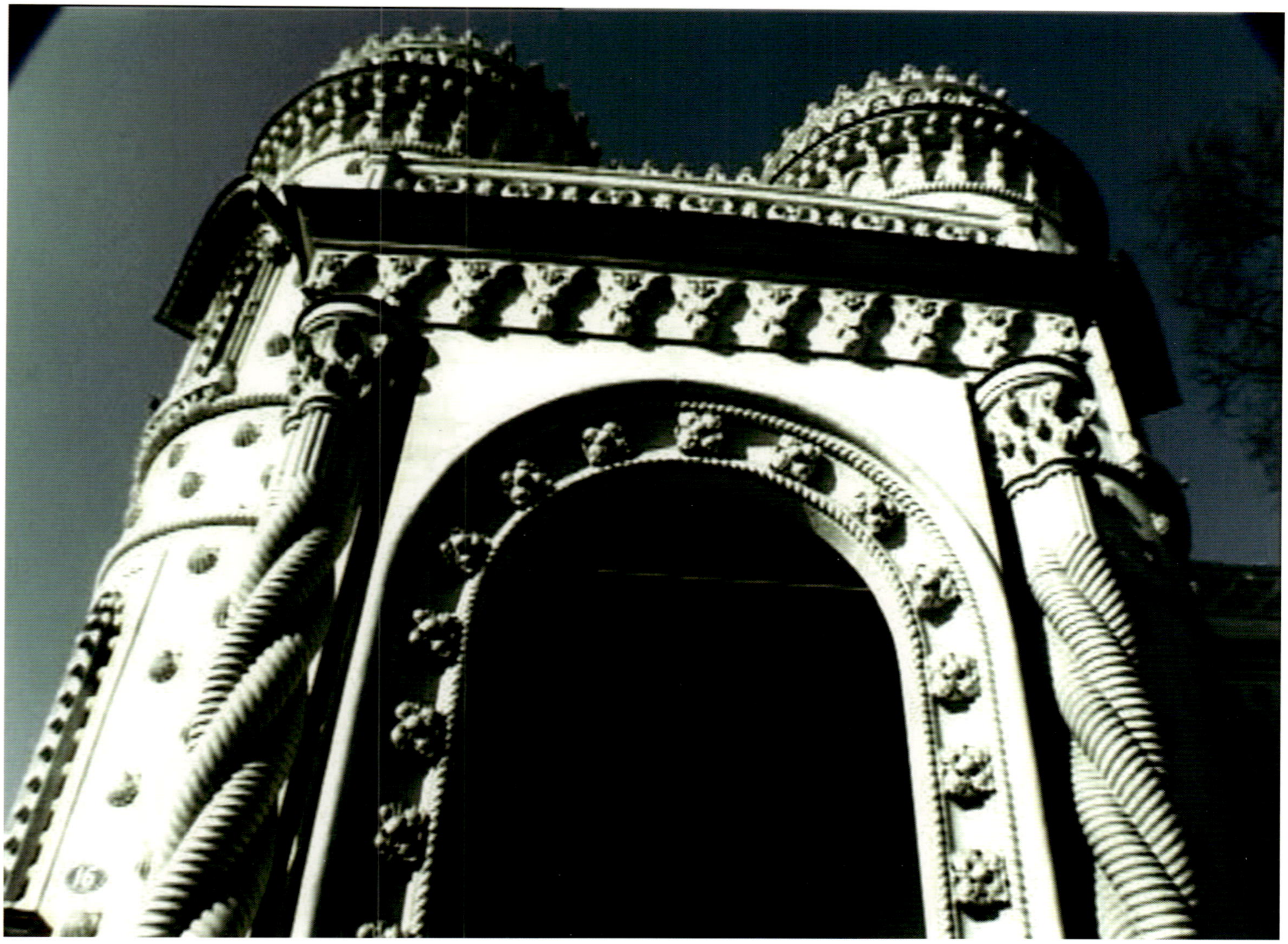

OTOLITH, 2003, stills

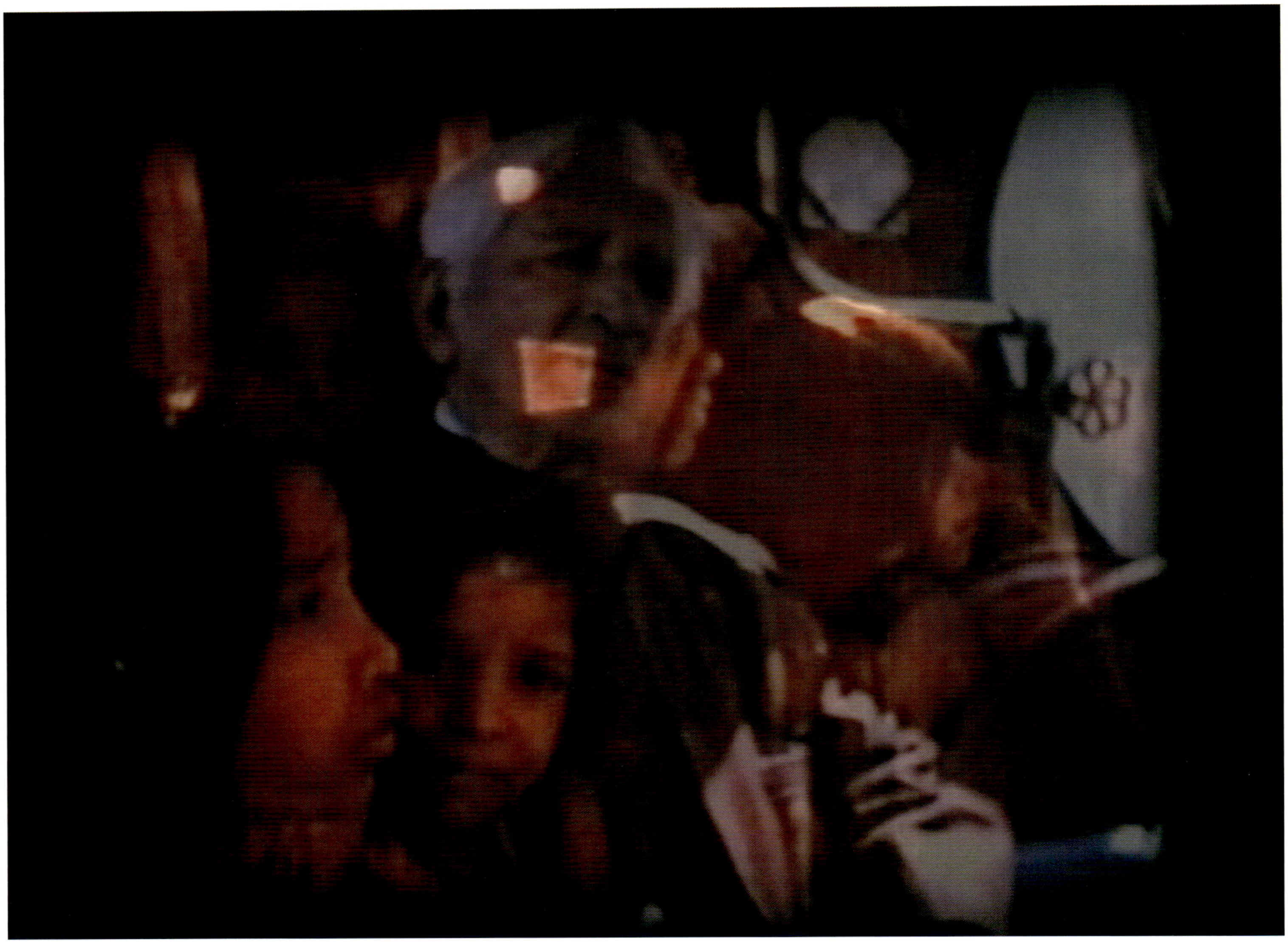

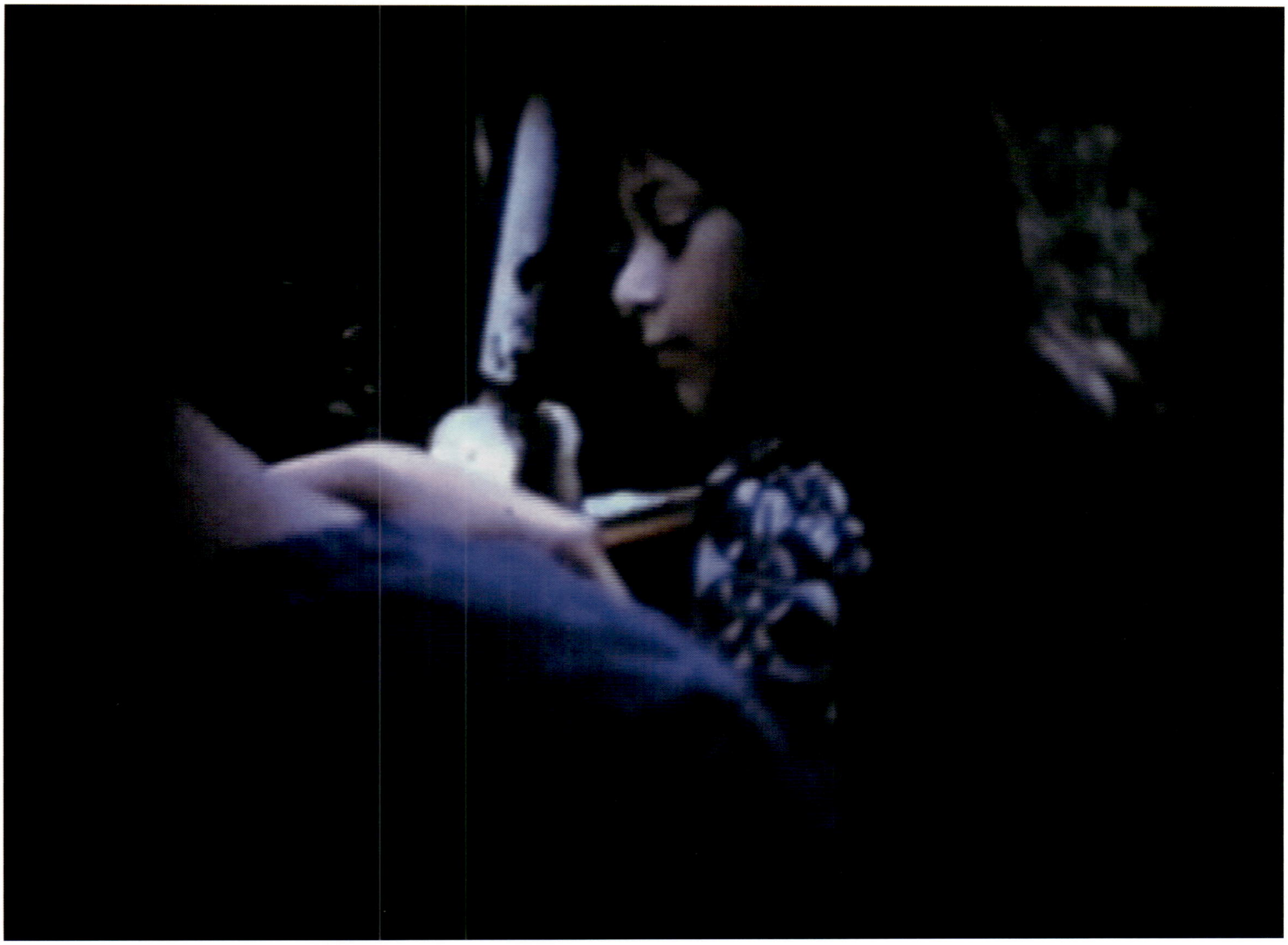

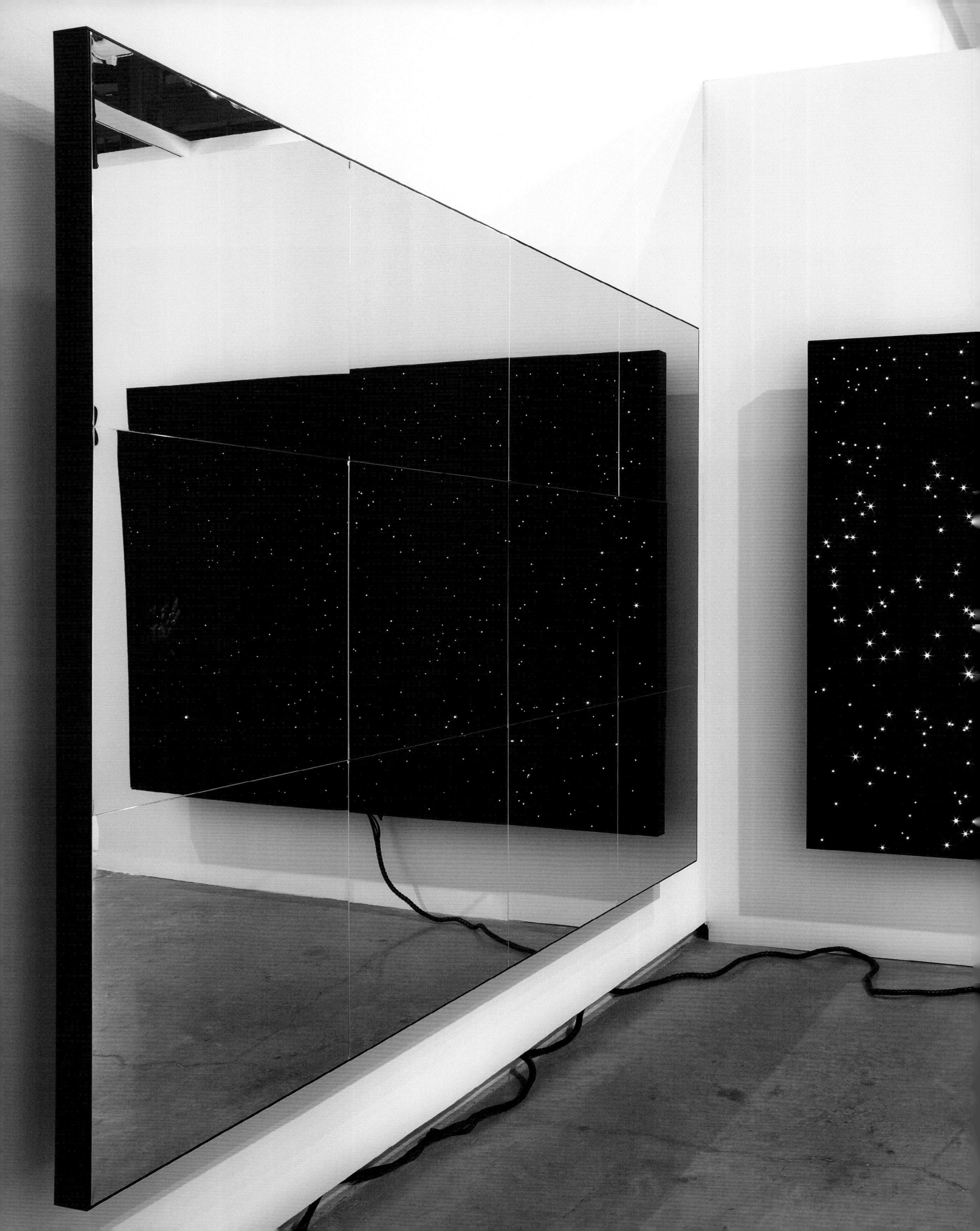

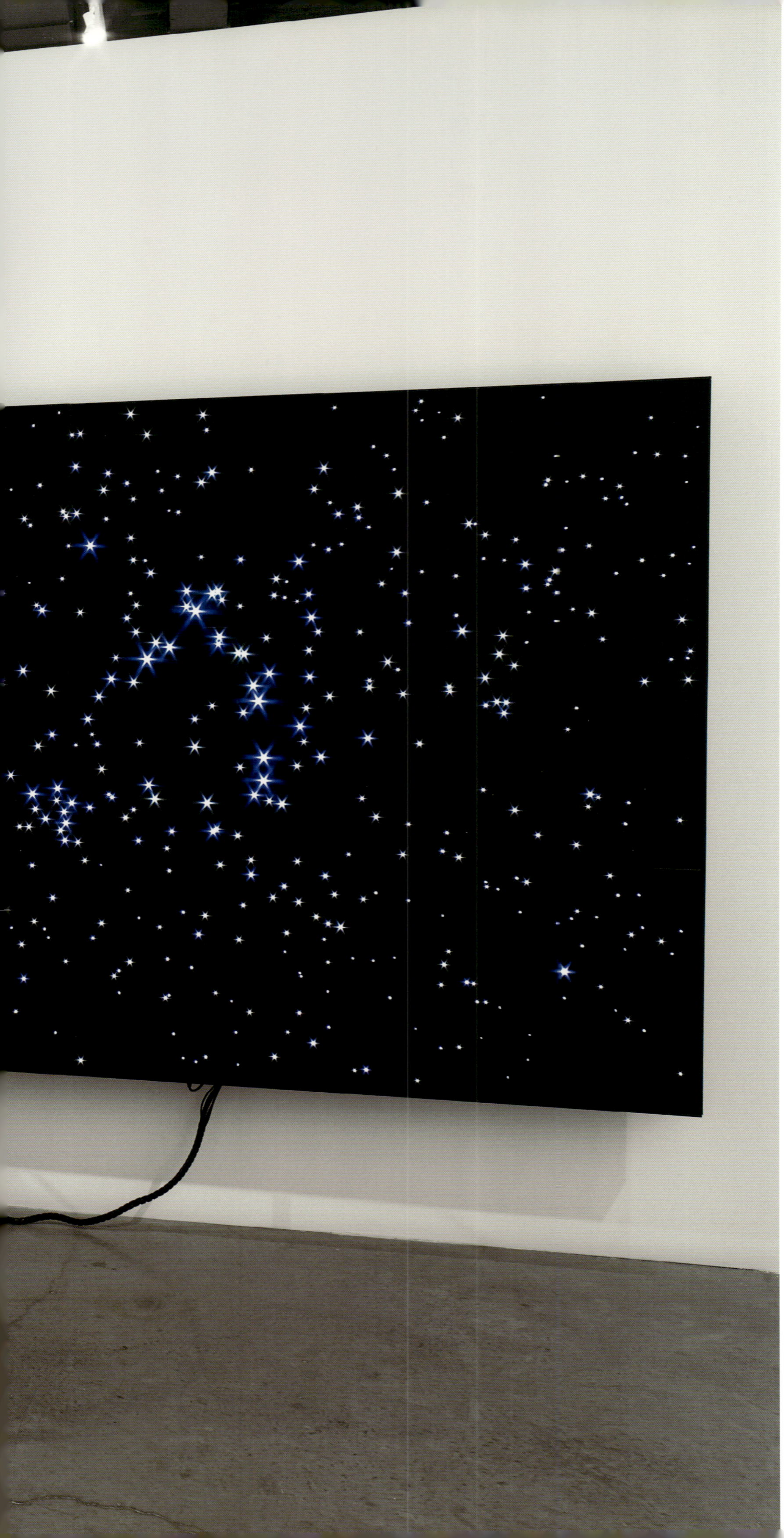

ANGELA BULLOCH

NIGHT SKY: MARS FROM VENUS.12, 2008
LEDs (light-emitting diodes), neoprene,
animated program, control gear, structural elements,
power suppliers, and various cables.
Courtesy Esther Schipper, Berlin

FRED TOMASELLI

CAR BOMB, 2008
Gouache on printed watercolour paper, 21 x 27.3 cm.
Collection Heidi L. Steiger, New York

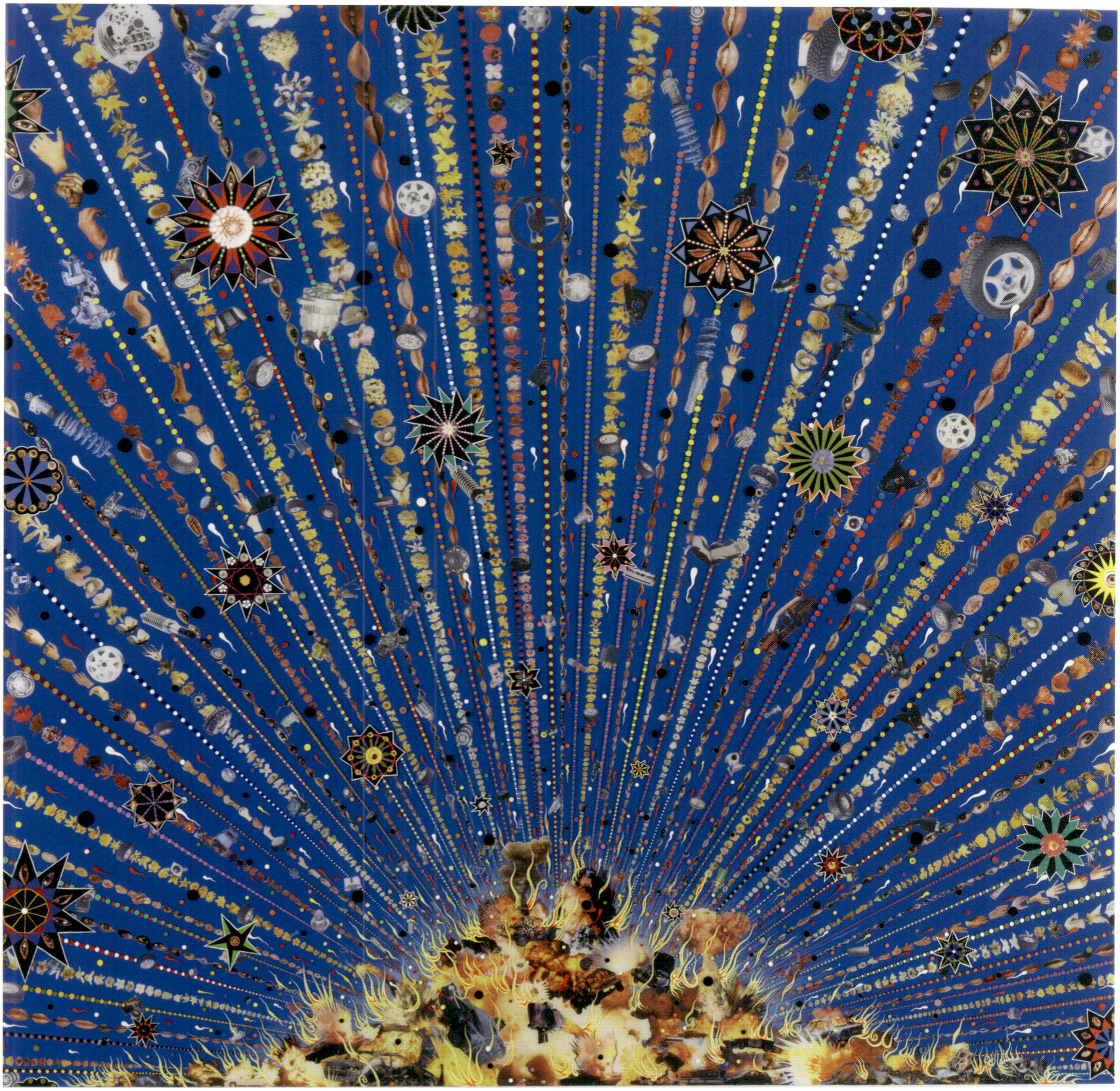

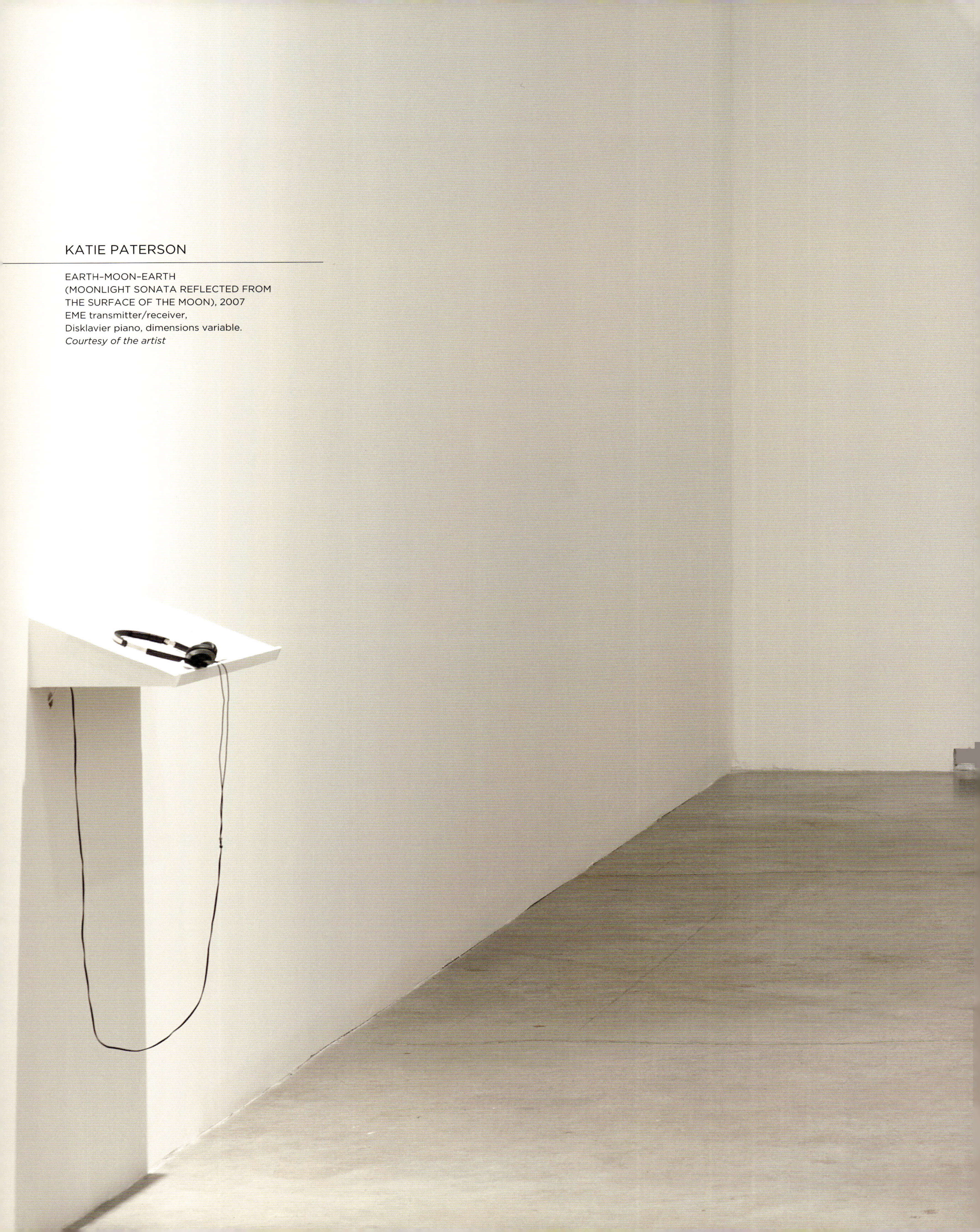

KATIE PATERSON

EARTH–MOON–EARTH
(MOONLIGHT SONATA REFLECTED FROM
THE SURFACE OF THE MOON), 2007
EME transmitter/receiver,
Disklavier piano, dimensions variable.
Courtesy of the artist

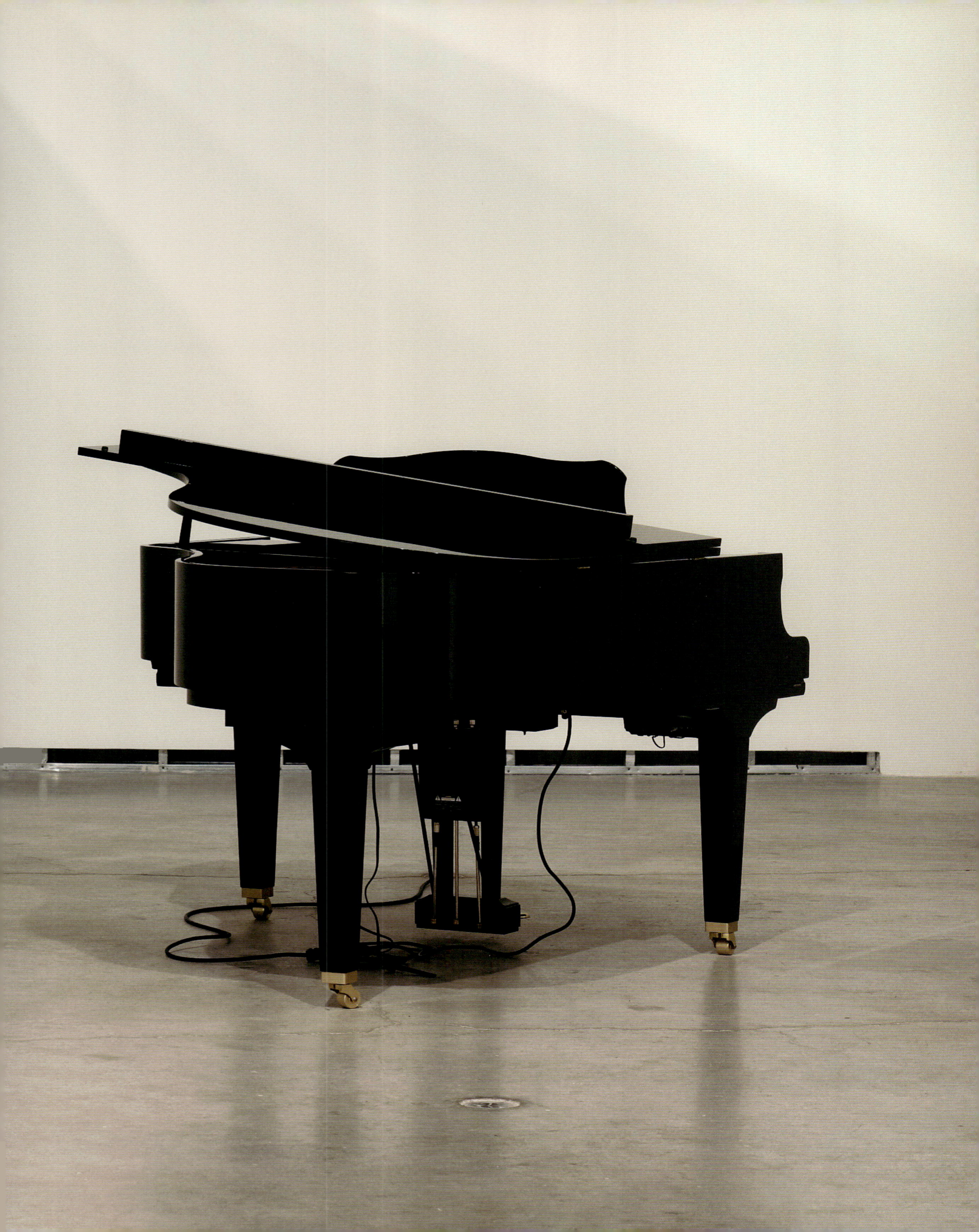

KATIE PATERSON

EARTH–MOON–EARTH
(MOONLIGHT SONATA REFLECTED FROM
THE SURFACE OF THE MOON), 2007

Earth–Moon

Moon–Earth

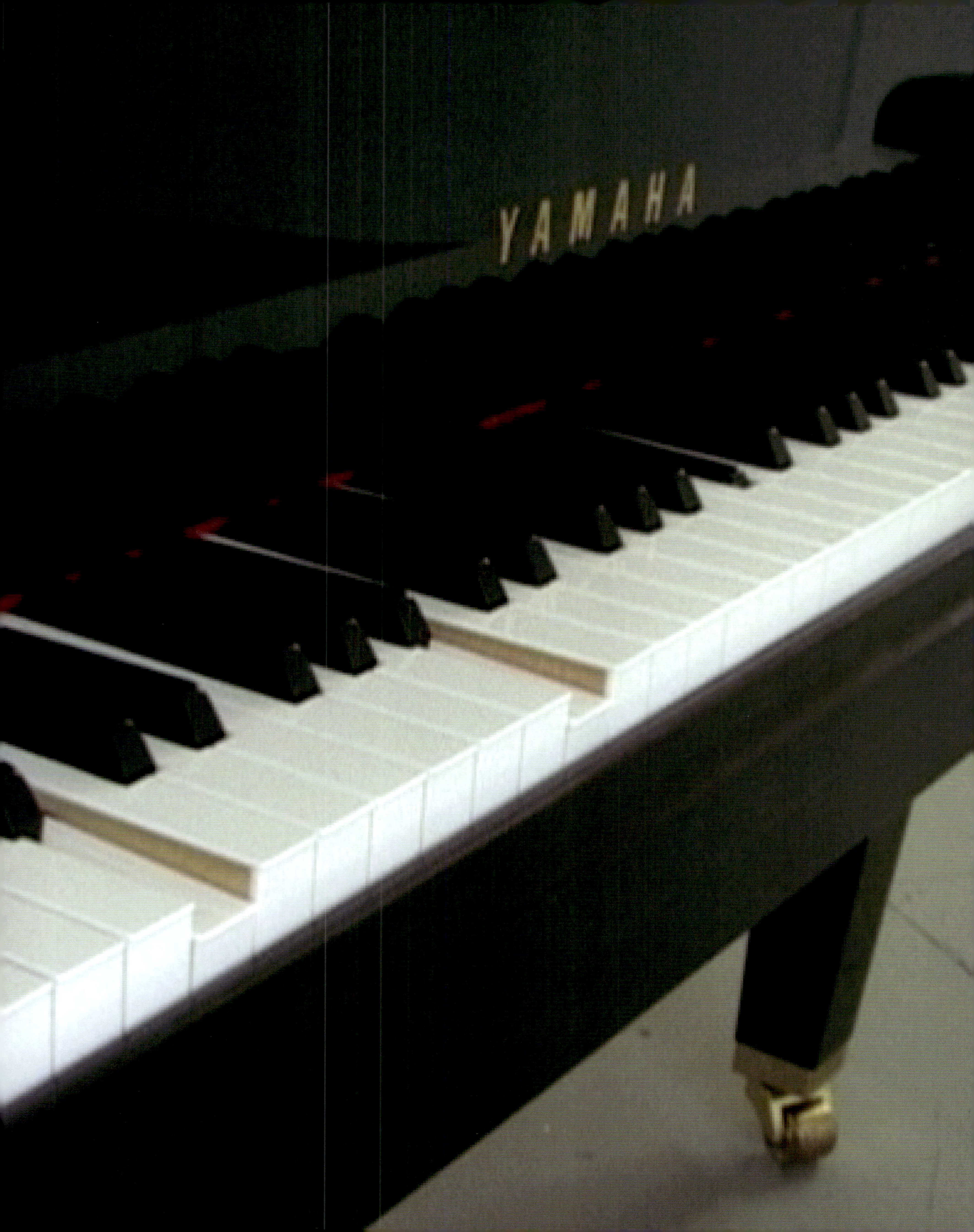
YAMAHA

ADRIAN PACI

PER SPECULUM, 2006
Printed photograph, 80 x 120 cm.
Courtesy Galleria Francesca Kaufmann, Milan

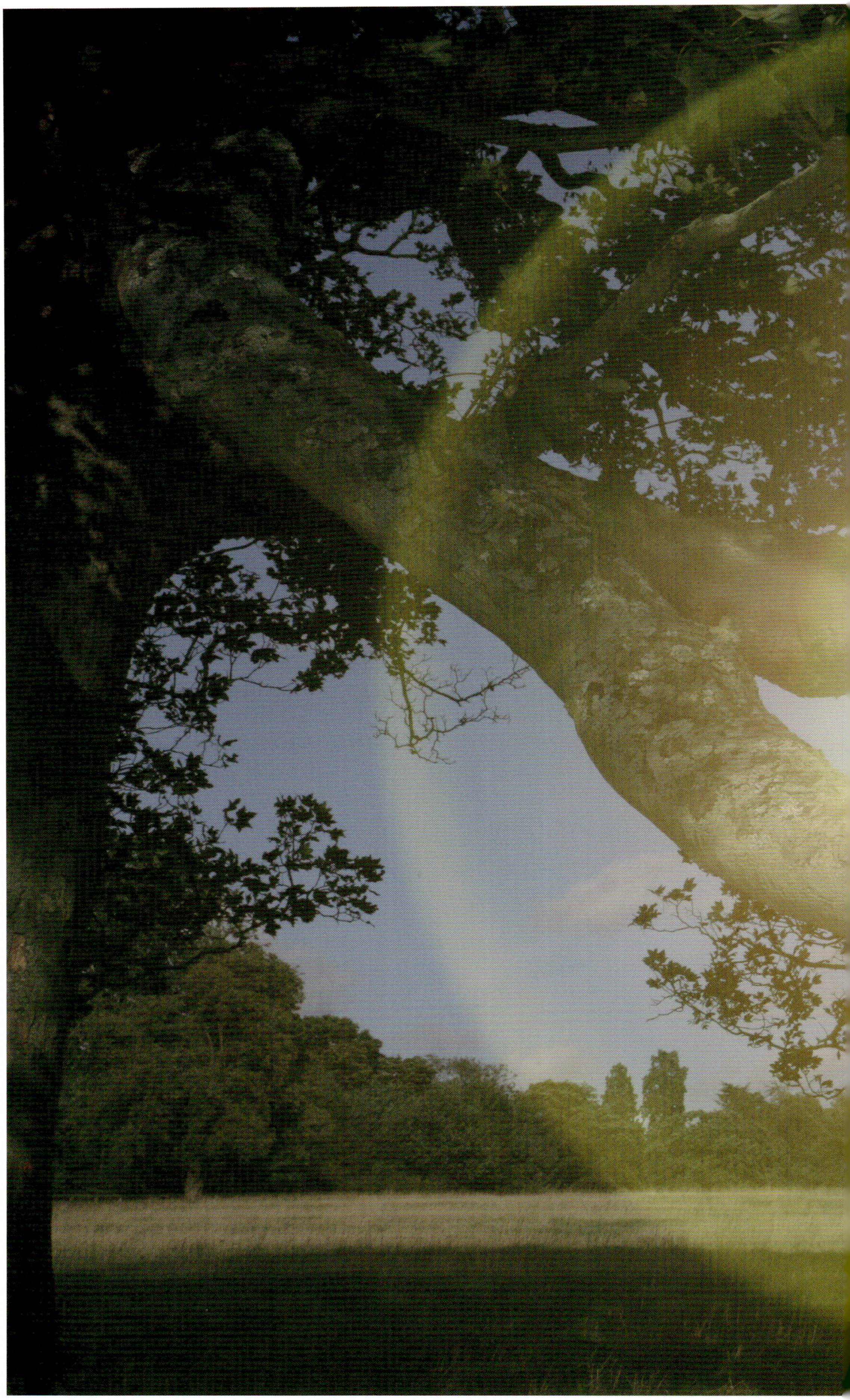

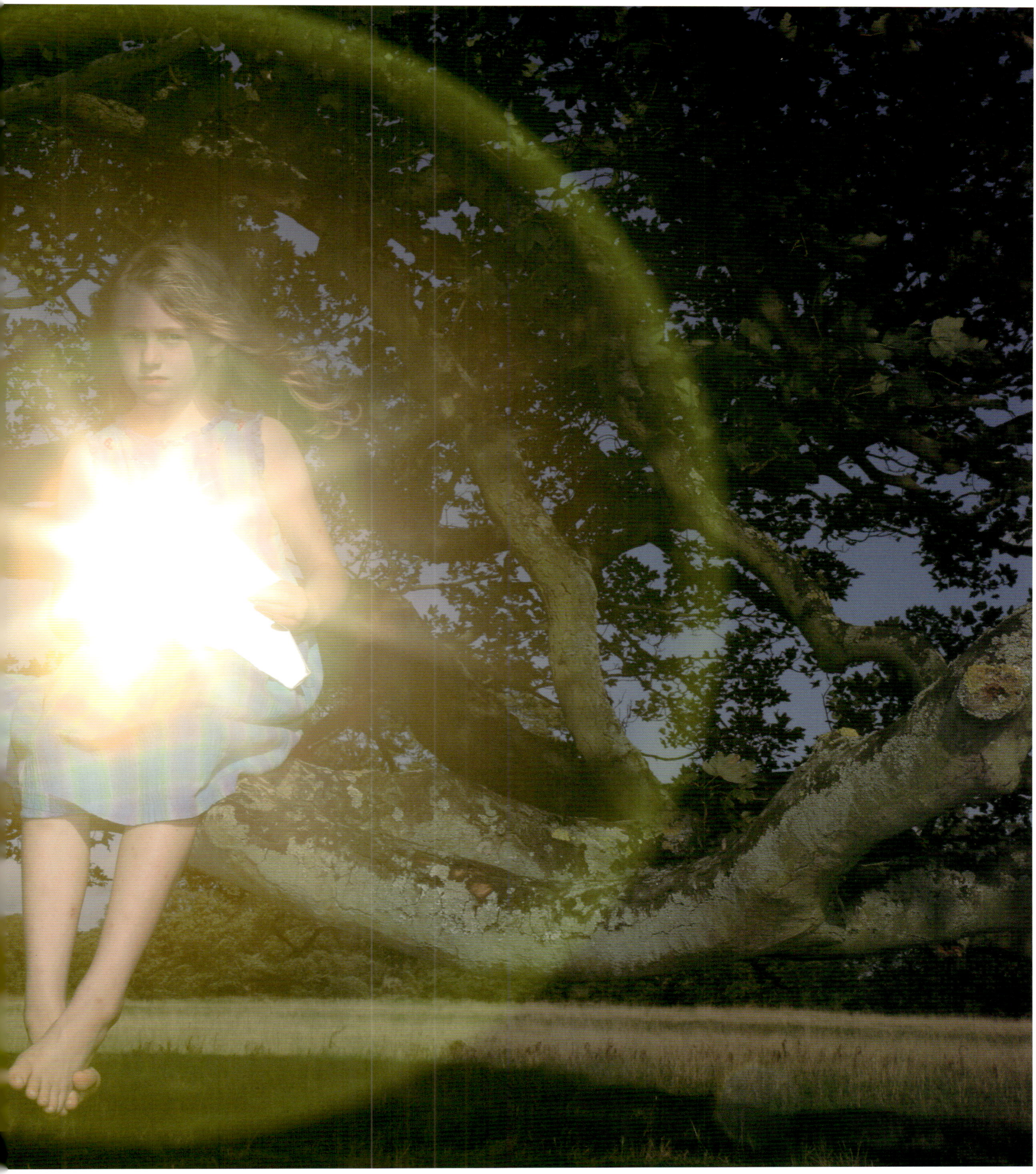

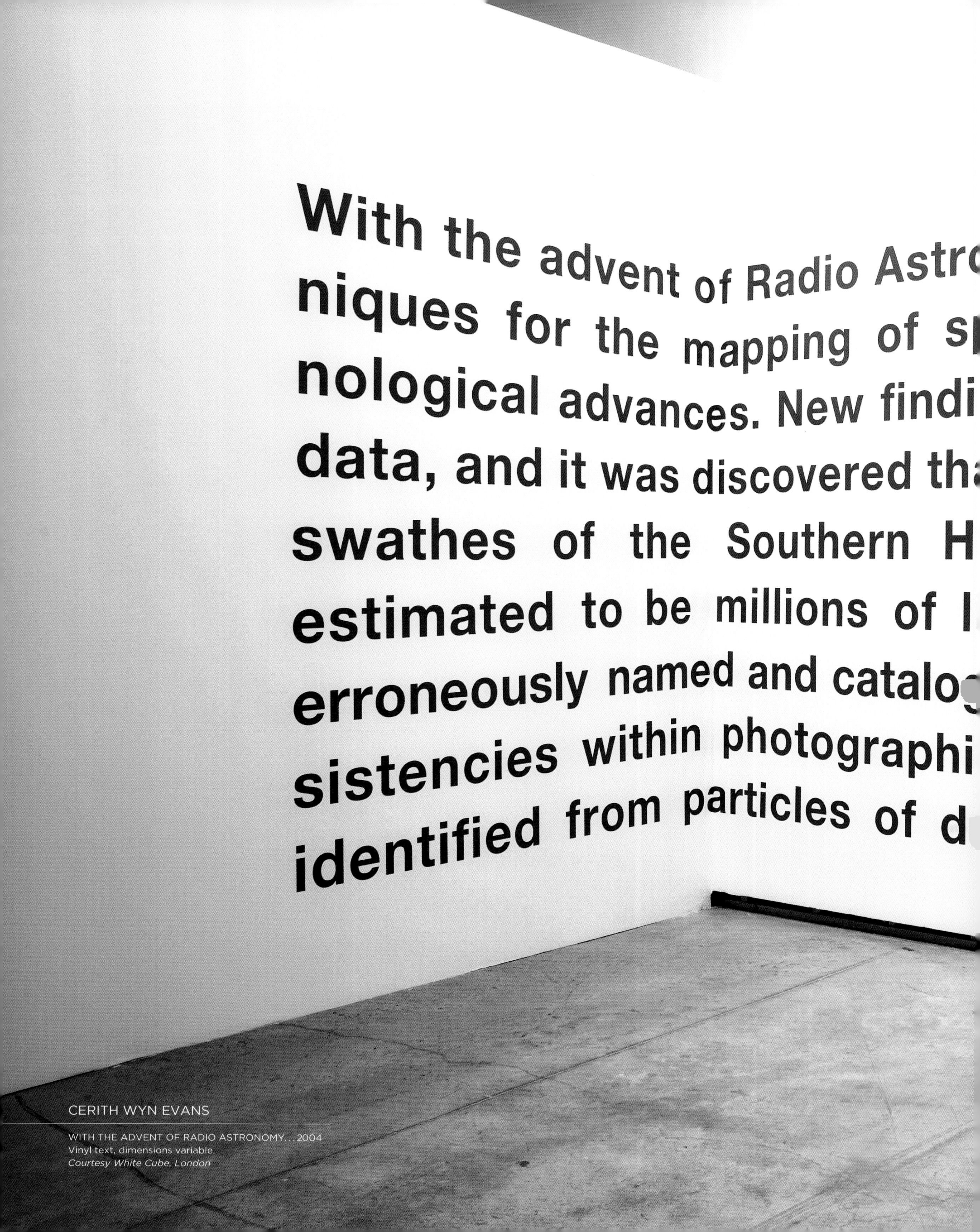

CERITH WYN EVANS

WITH THE ADVENT OF RADIO ASTRONOMY... 2004
Vinyl text, dimensions variable.
Courtesy White Cube, London

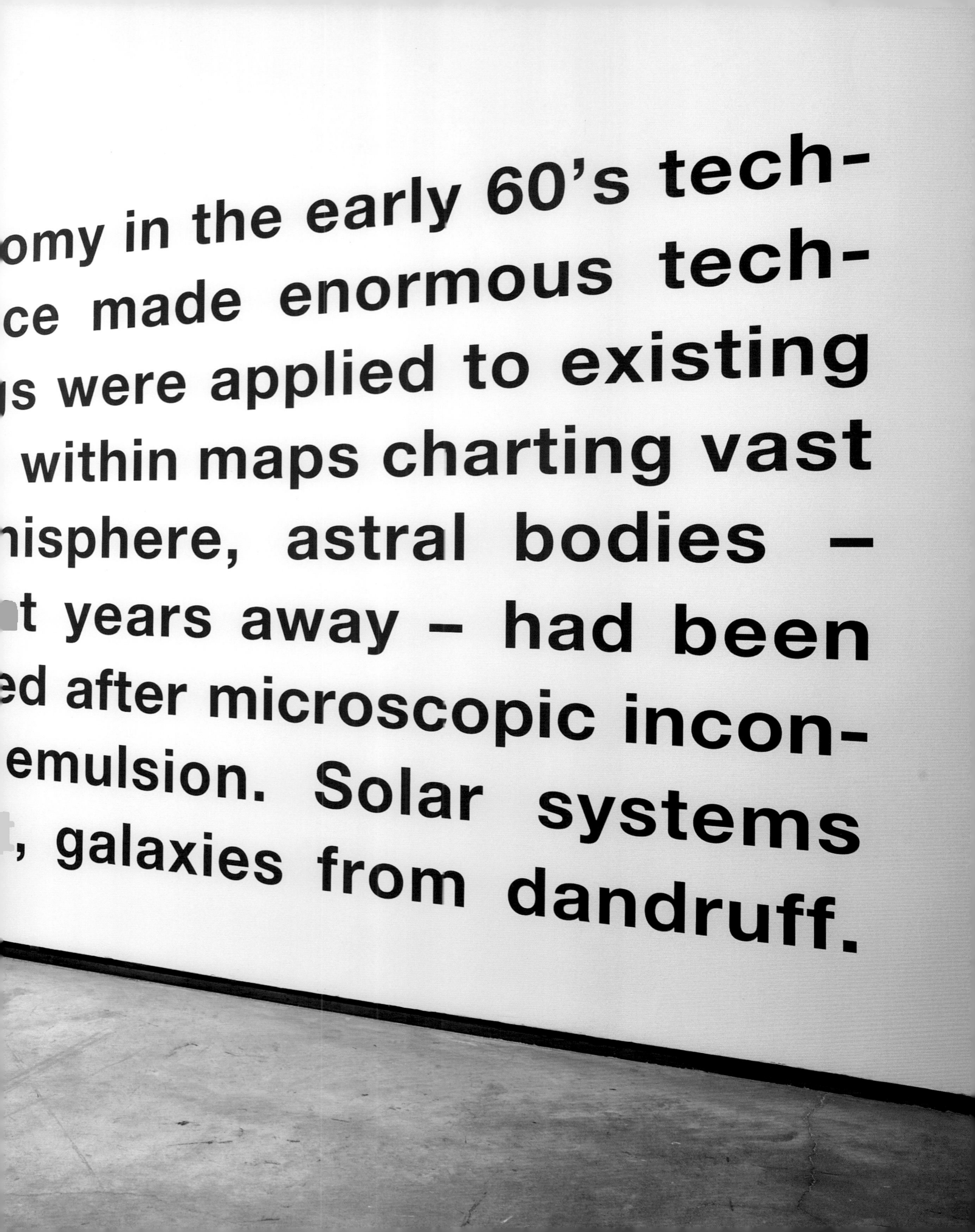
omy in the early 60's tech-
ce made enormous tech-
s were applied to existing
within maps charting vast
isphere, astral bodies –
t years away – had been
ed after microscopic incon-
emulsion. Solar systems
, galaxies from dandruff.

MIRCEA CANTOR

AIRPLANES AND ANGELS, 2008
Hand-woven wool carpet, 198 x 300 cm.
Courtesy of the artist and Yvon Lambert, Paris/New York.

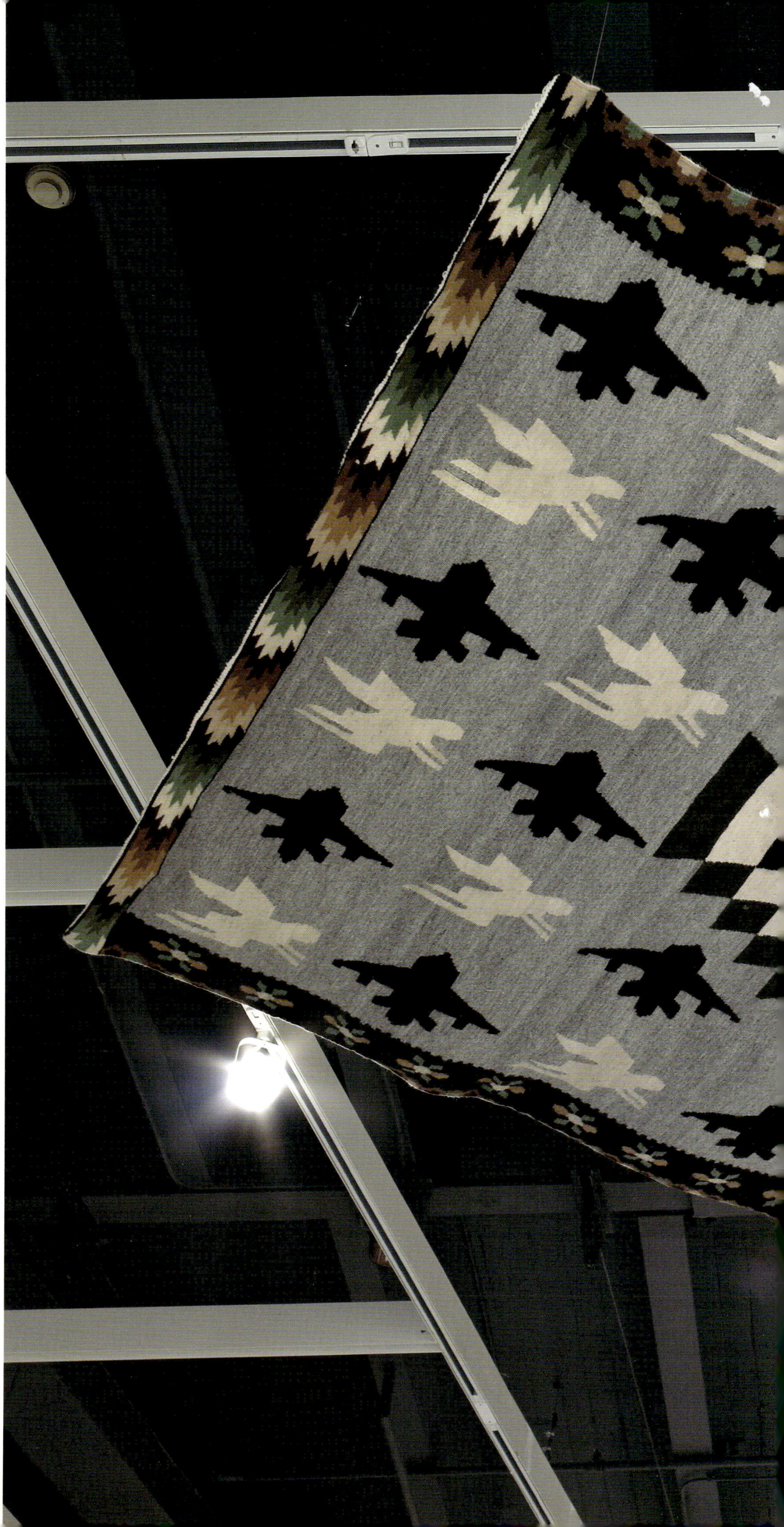

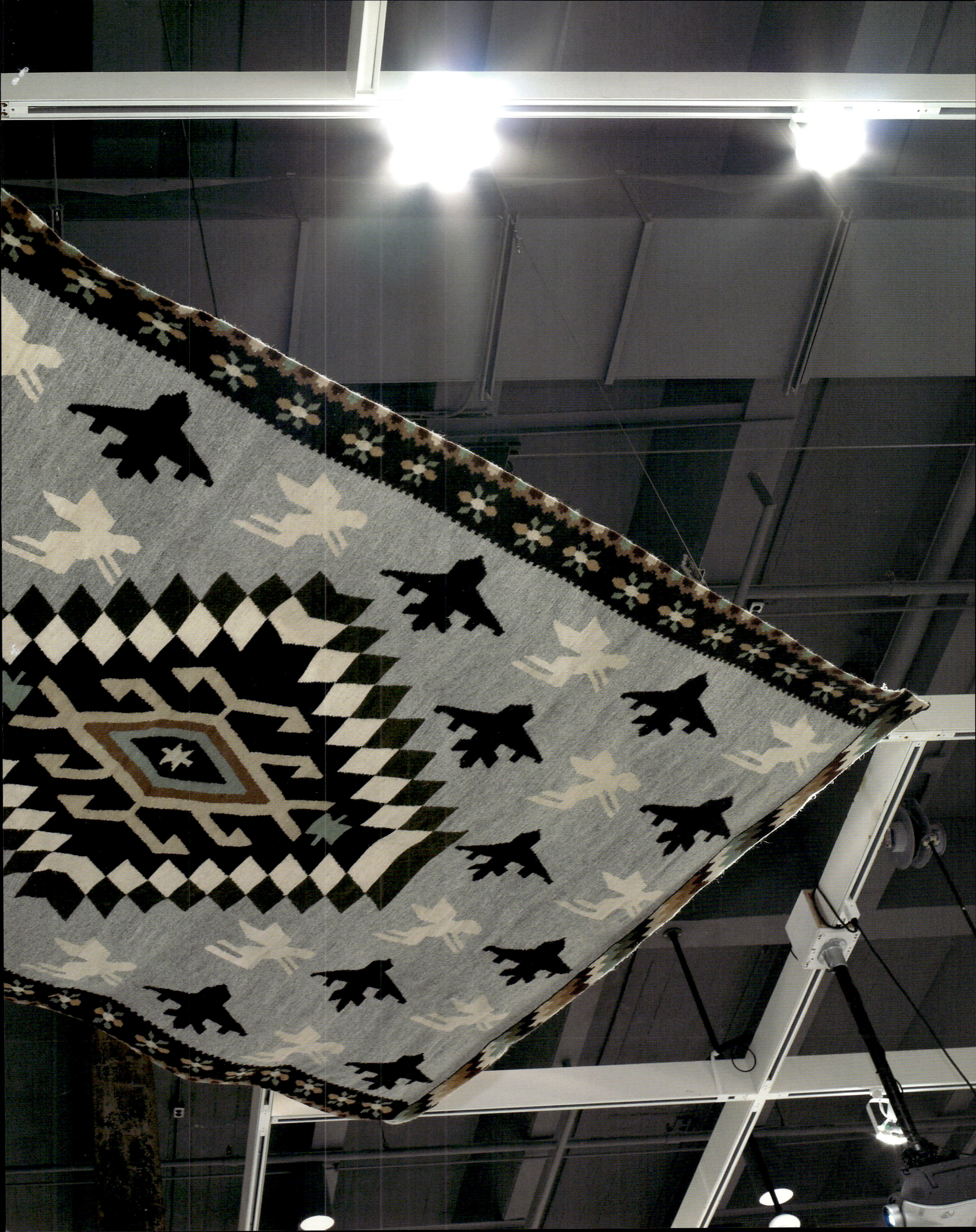

JANET CARDIFF & GEORGE BURES MILLER

TELEPHONE / TIME, 2004
Desk, chair, lamp, telephone, audio installation, 2 min., loop.
Collection Albright-Knox, Buffalo

Man: So geographically there will be a variation in time between Sweden and Vancouver due to the different distances from the center of the earth – from the center of gravity. Janet: Right. Man: So one would actually get a measurable difference in time. Janet: At one point you were talking to me about the multiple time periods that appear at the edge of the event horizon. Man: At the event horizon: at the edge of physics in some sense. A person wouldn't experience time in an ordinary sense. You couldn't be talking about doing things. There wouldn't be intervals between events most likely rather there would be transformations going on. But not ones that a person could linearly track. Janet: So the person would be in multi-dimensions at the same time? Man: In some sense. Janet: Or if you have a shock. Like one time I was walking down the street and someone ran by and yelled at the same time as they passed my ear and everything completely slowed down for me. Man: That makes entire sense. That your sense of time would be changing in relation to the world around you and how time progresses for you would change as those relationships alter. So in that instance the circumstances you are in change abruptly. Janet: Right. Man: Your sense of time alters and one see the shift in the sense of time and one can observe that because its such a rapid shift rather than being a fluent gradual shift which would be more common in our daily lives. Our sense of time is changing very gradually so we don't tend to notice the changing. Janet: So if everybody is working on his own sense of time then how can you have any consistent time? That isn't logical. Man: That's right. Janet: And that would mean that than there would be a multi-dimensionality going on constantly . . . Man: Exactly.

Repeats from the top.

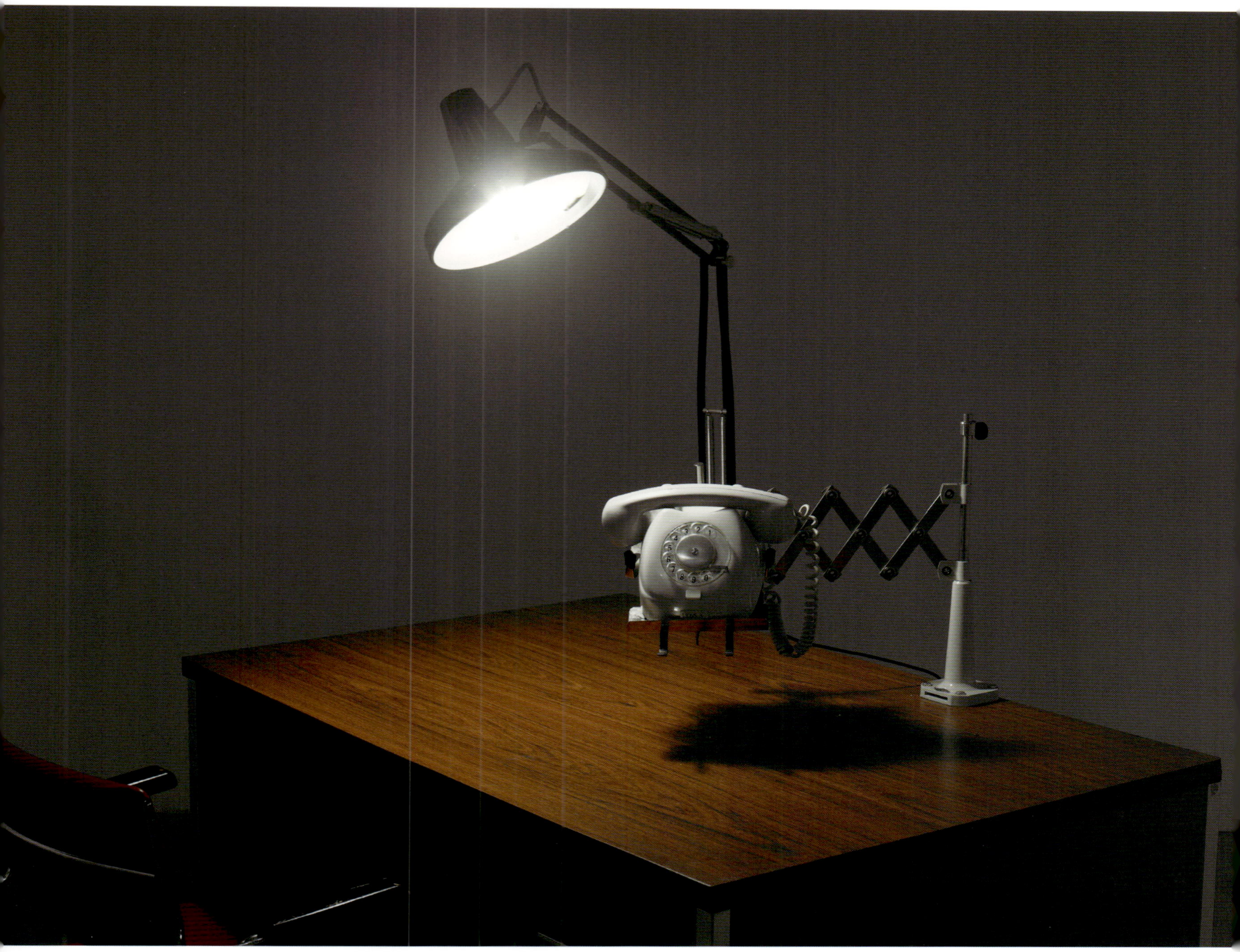

ANN VERONICA JANSSENS

SIDE (STUDIO VERSION), 2006, still
Video projection, 3 min., 3 sec., loop, colour, silent.
Courtesy Galerie Micheline Szwajcer, Antwerp

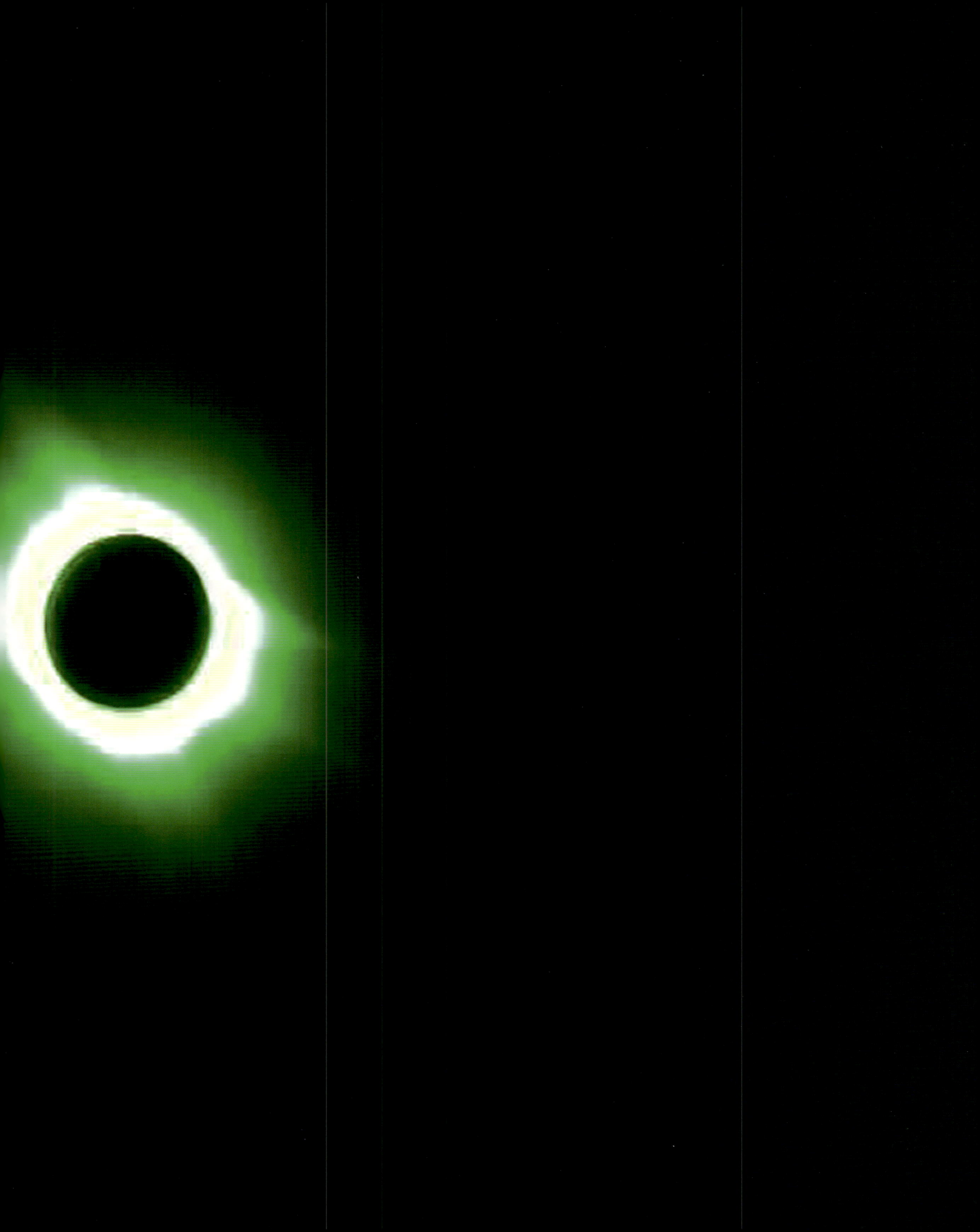

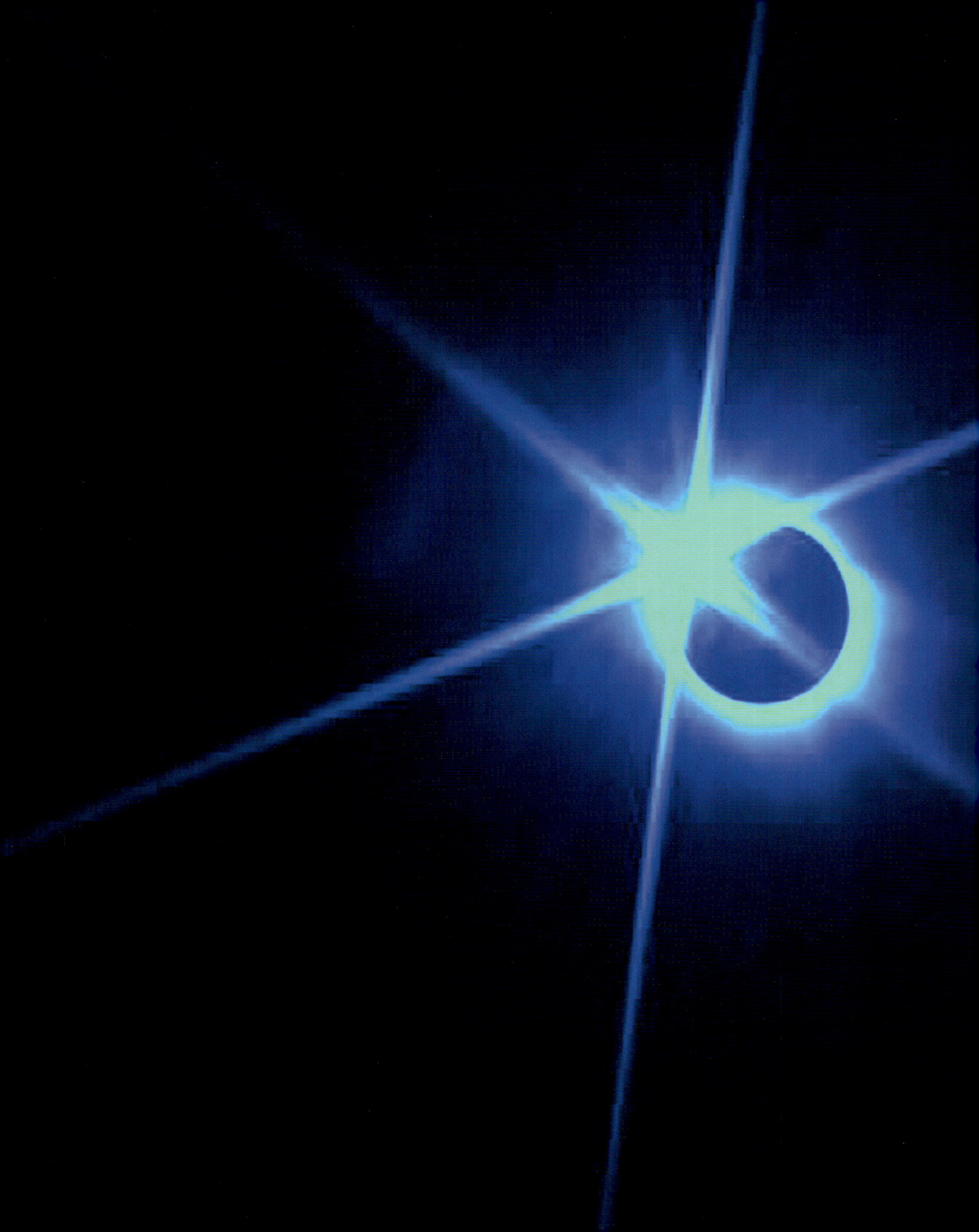

ANN VERONICA JANSSENS

SIDE (STUDIO VERSION), 2006, still

ANN VERONICA JANSSENS

SIDE (STUDIO VERSION), 2006, still

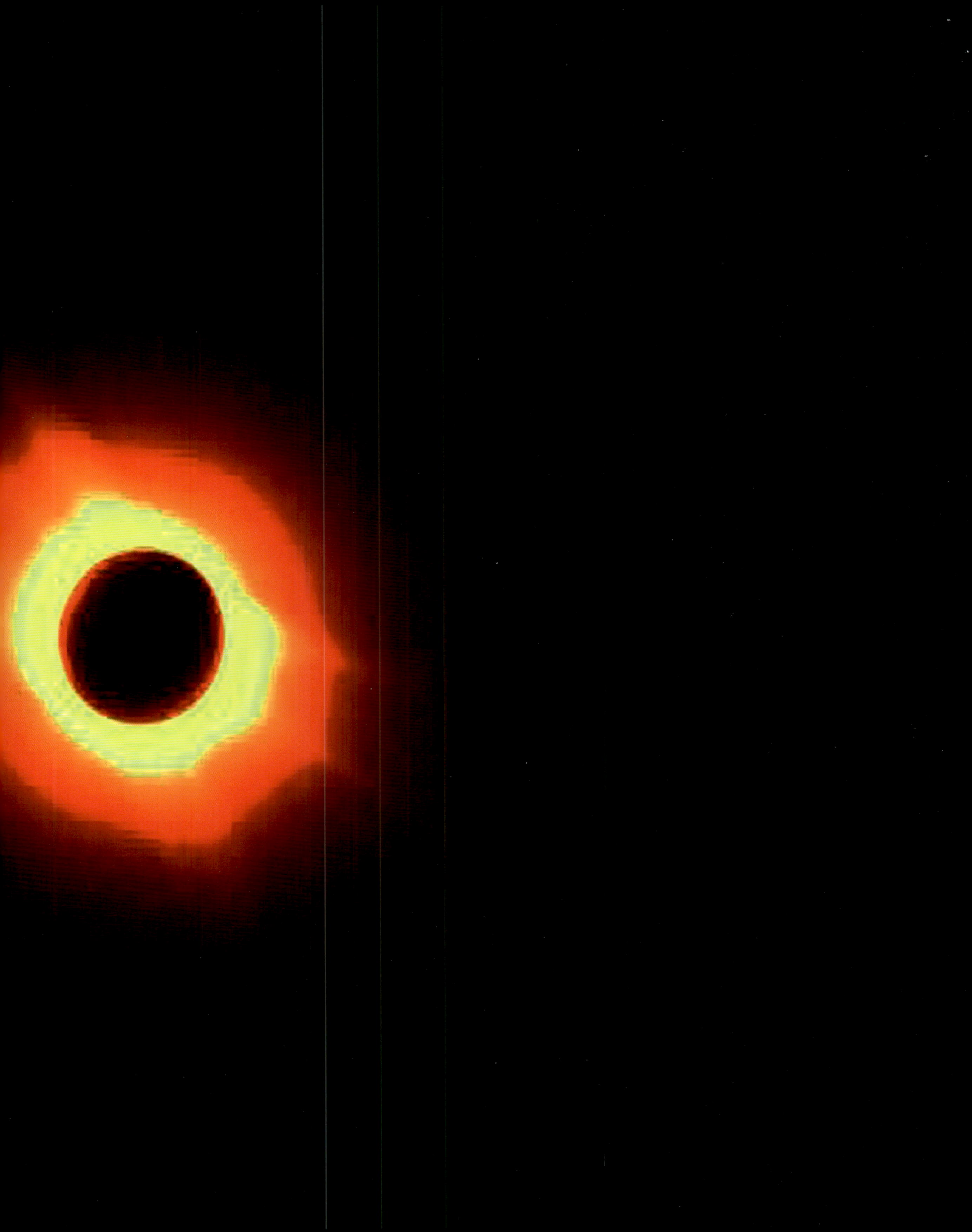

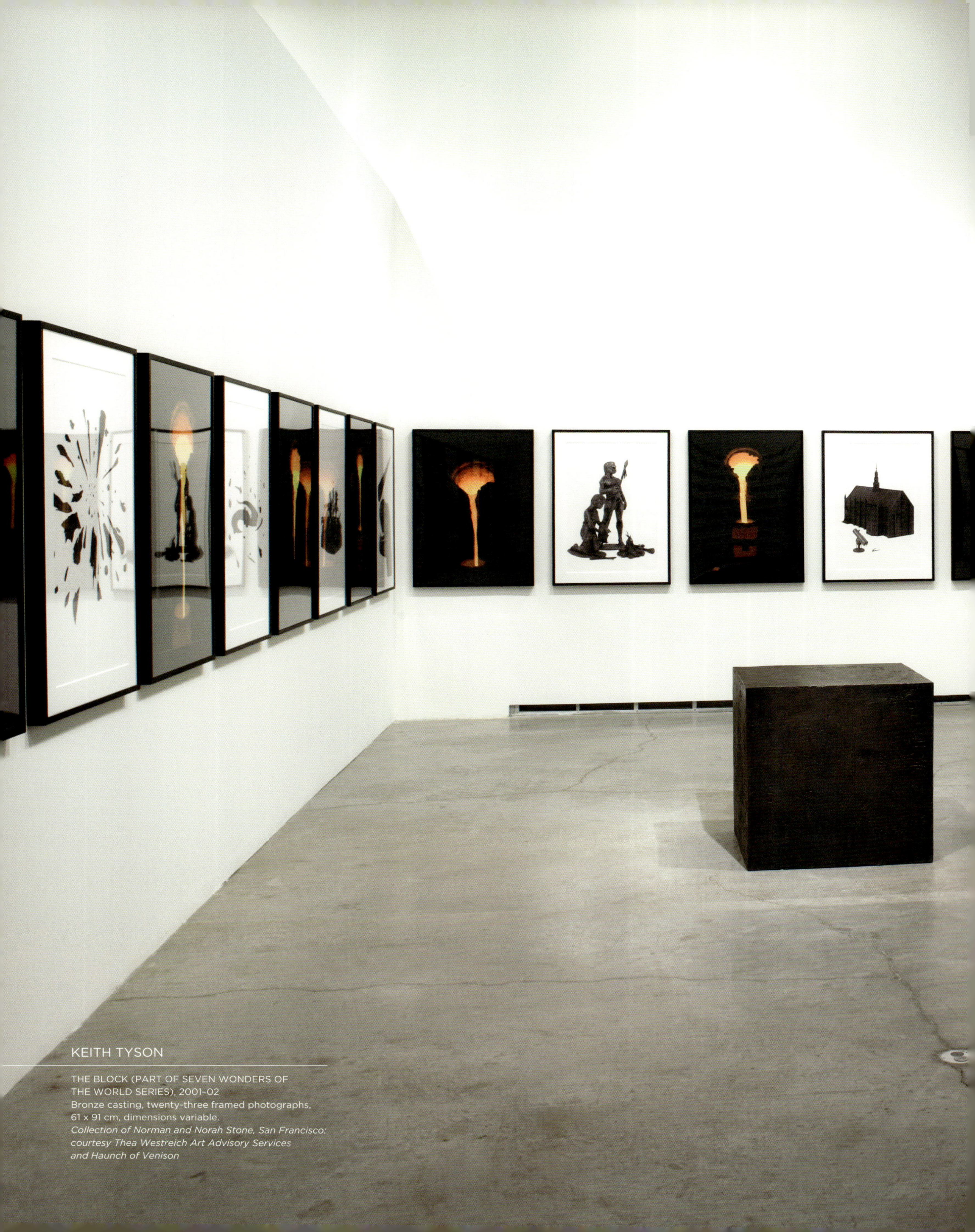

KEITH TYSON

THE BLOCK (PART OF SEVEN WONDERS OF THE WORLD SERIES), 2001–02
Bronze casting, twenty-three framed photographs, 61 x 91 cm, dimensions variable.
Collection of Norman and Norah Stone, San Francisco; courtesy Thea Westreich Art Advisory Services and Haunch of Venison

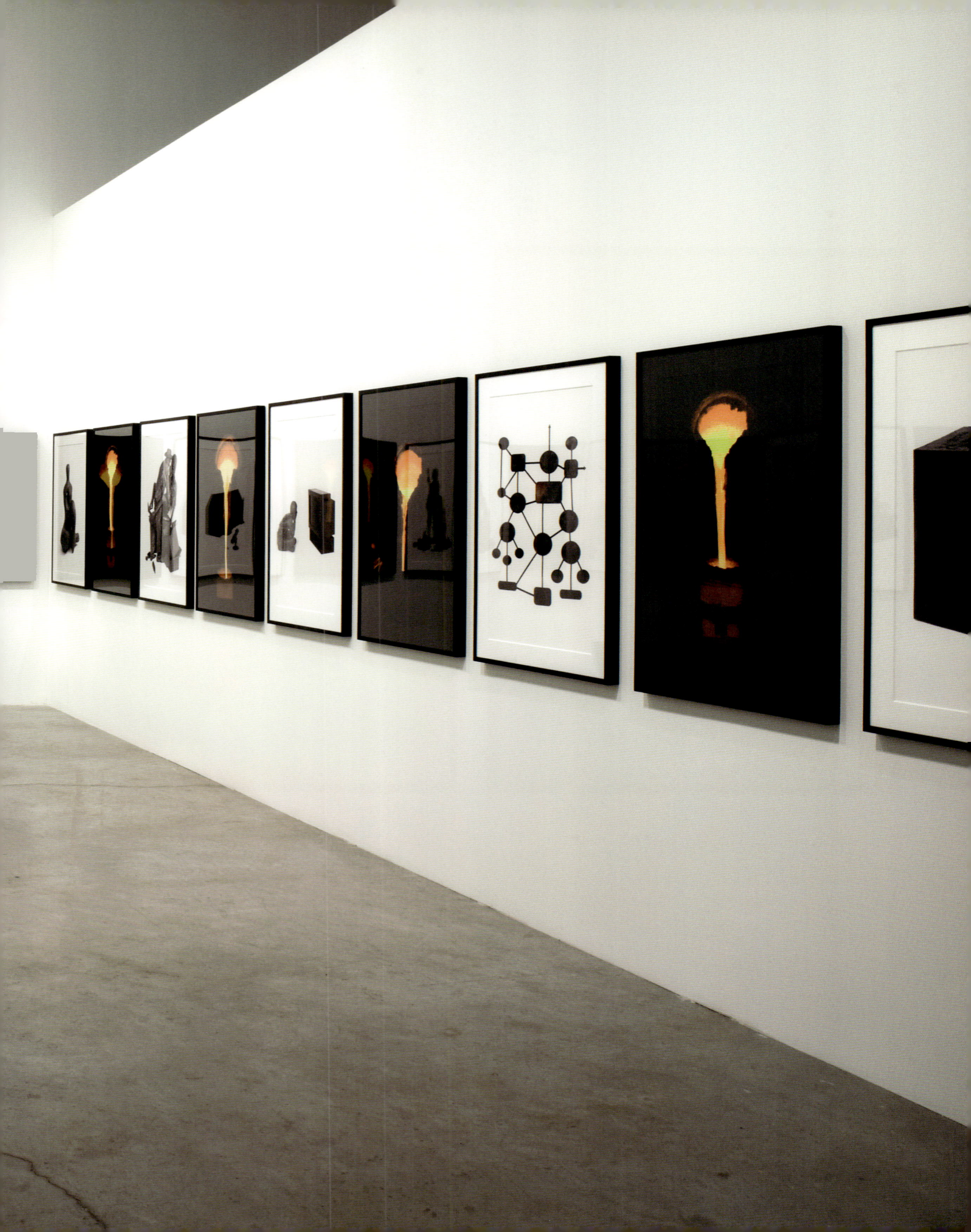

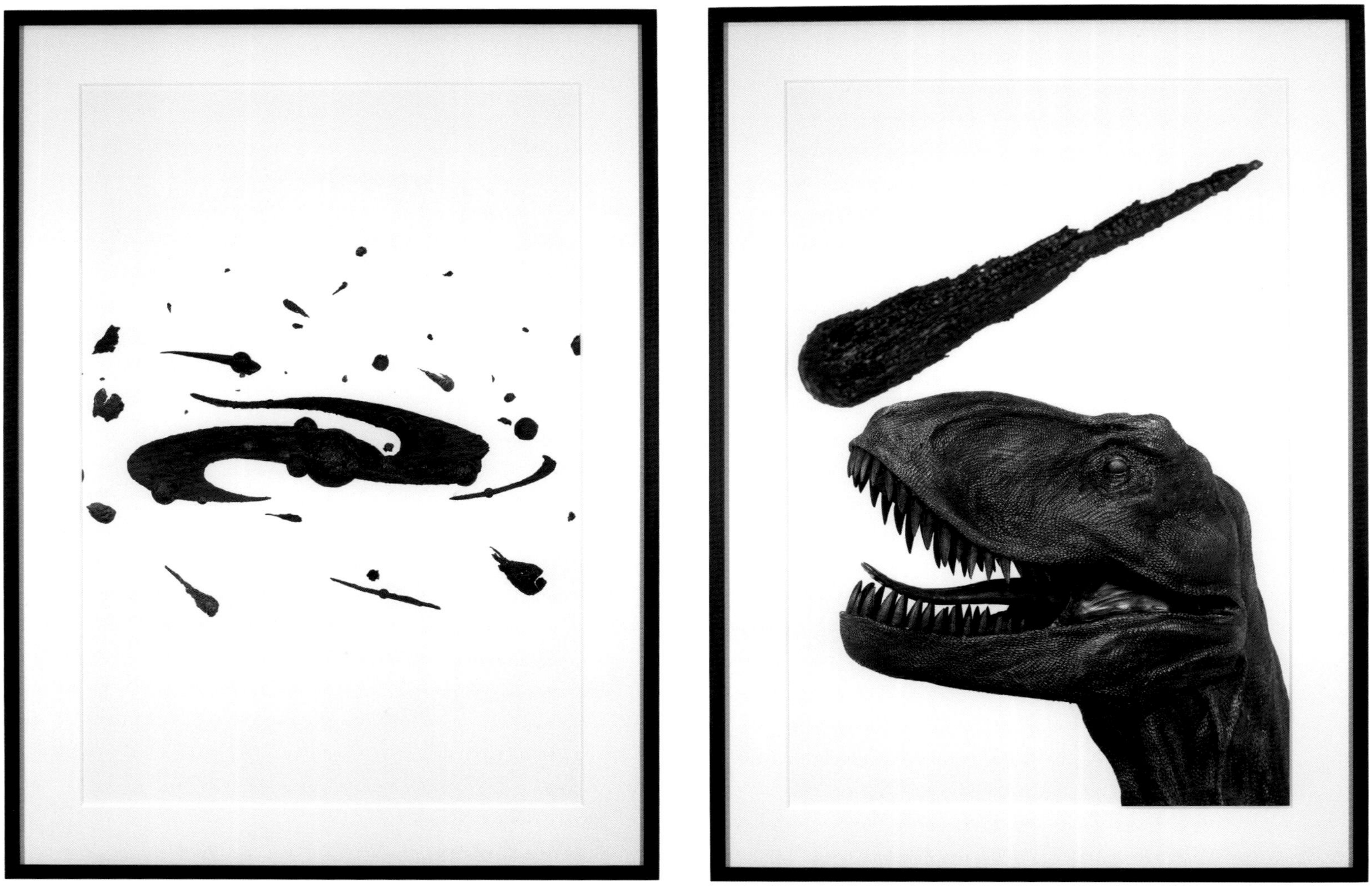

KEITH TYSON

THE BLOCK (PART OF SEVEN WONDERS OF THE WORLD SERIES), 2001–02, details

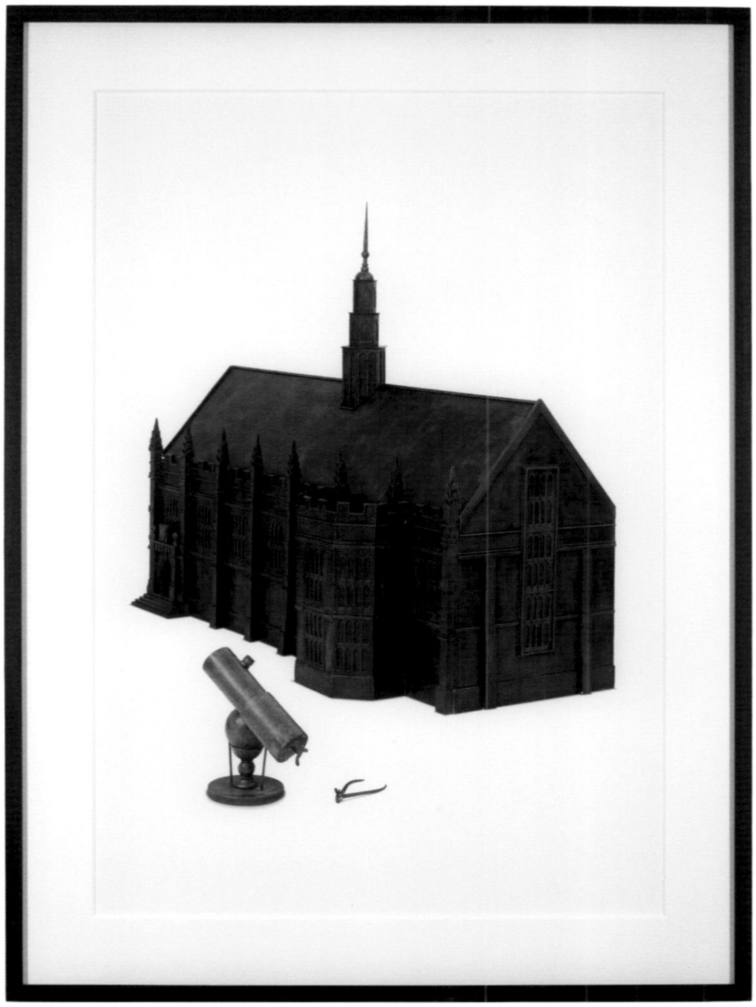

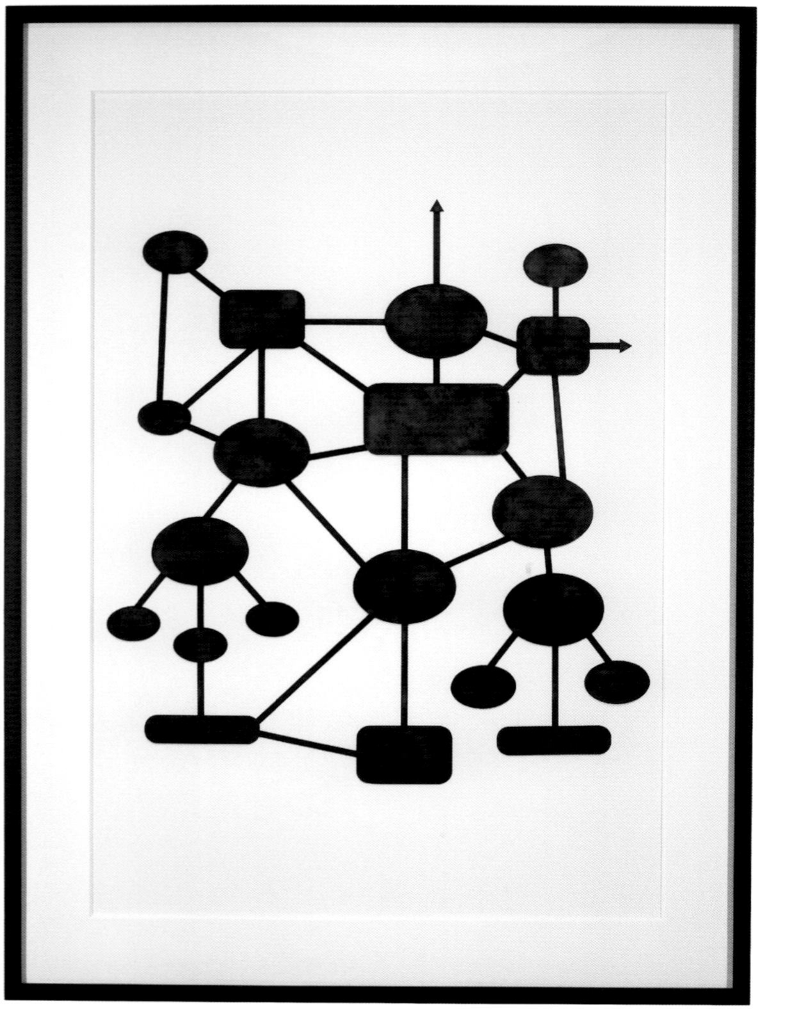

FRANZ ACKERMANN

ROTER PLATZ, 2007
Wall painting 350 x 300 cm, drawing
on aludibond 158 x 159 x 2.5 cm.
Collection Jay Smith and Laura Rapp

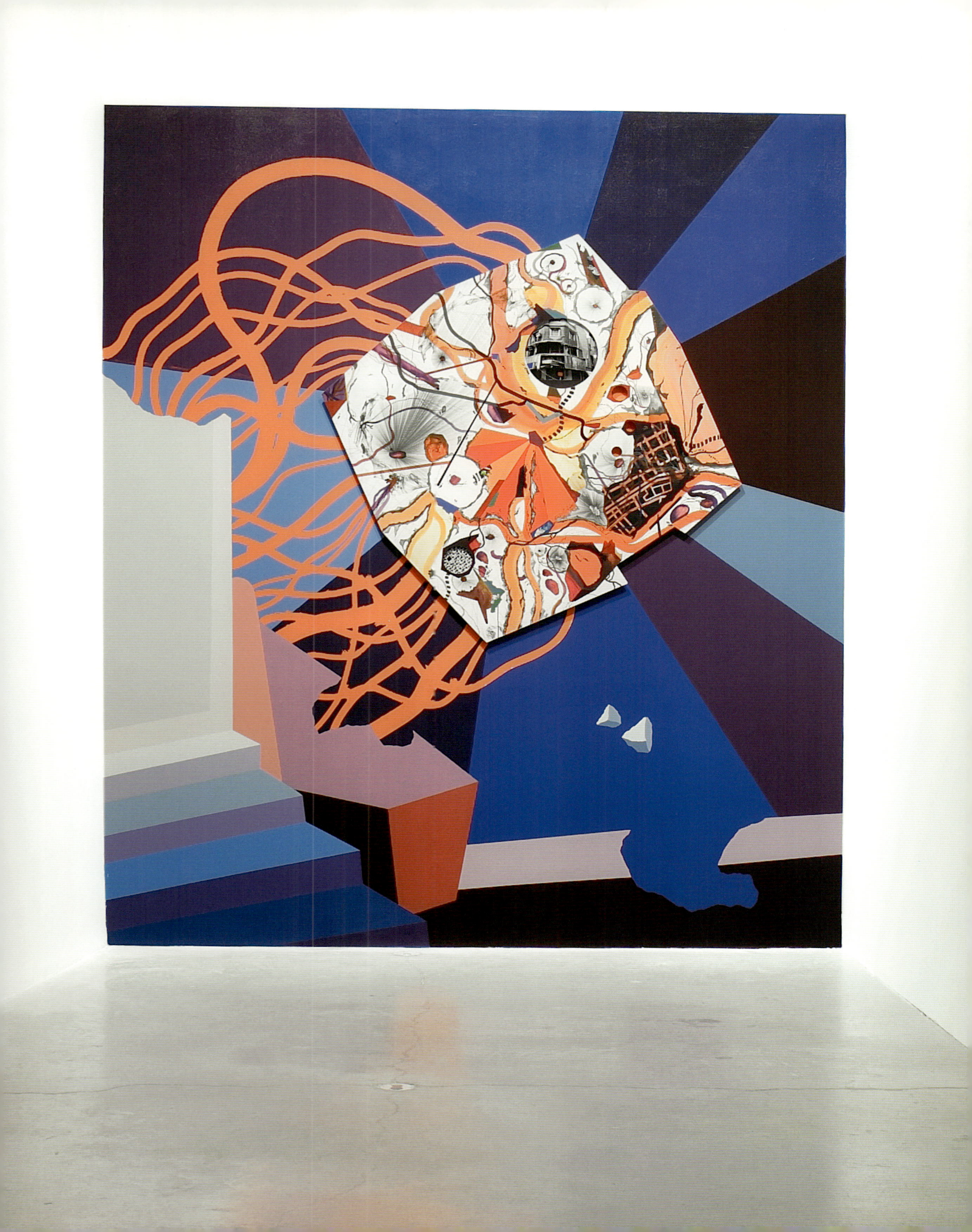

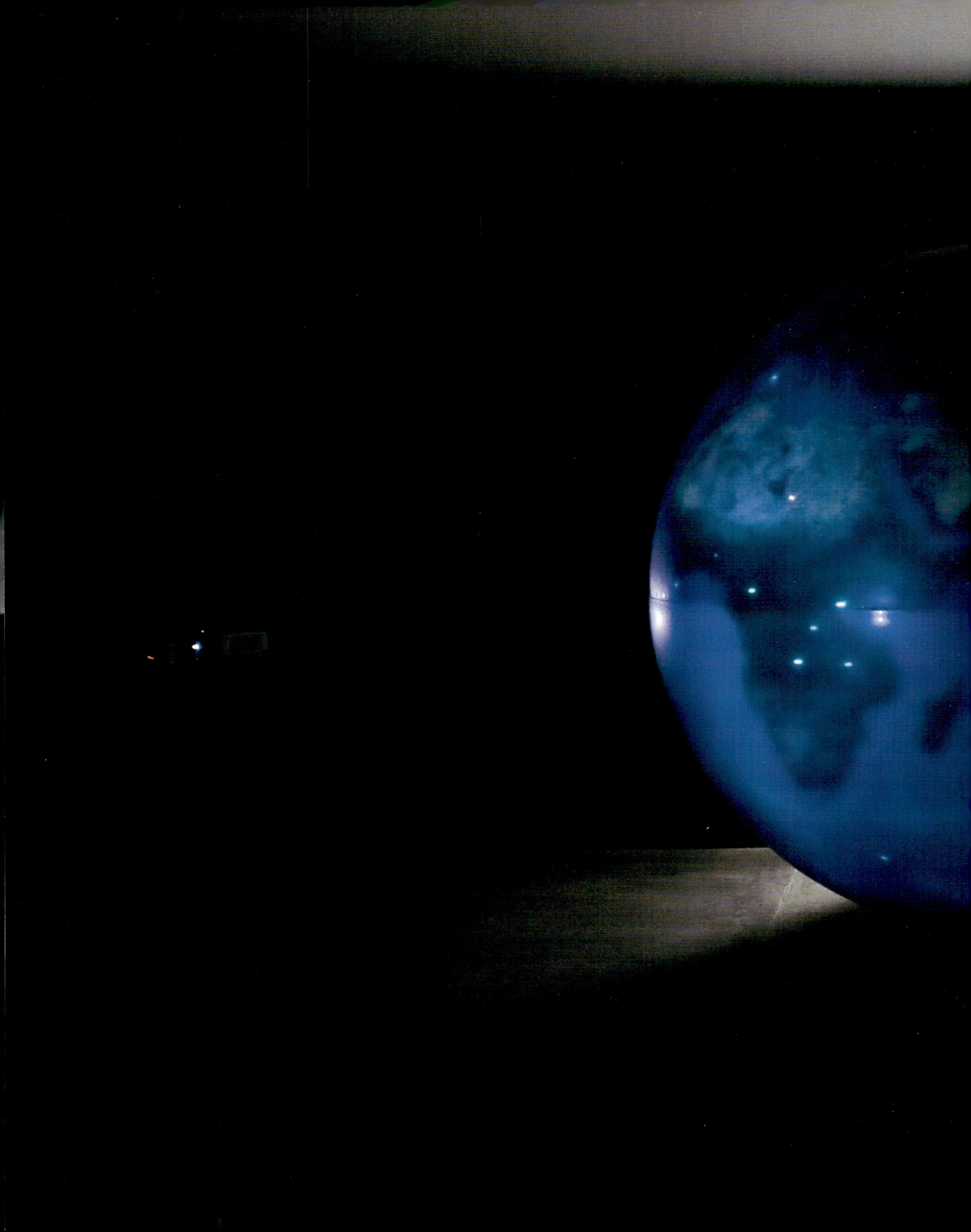

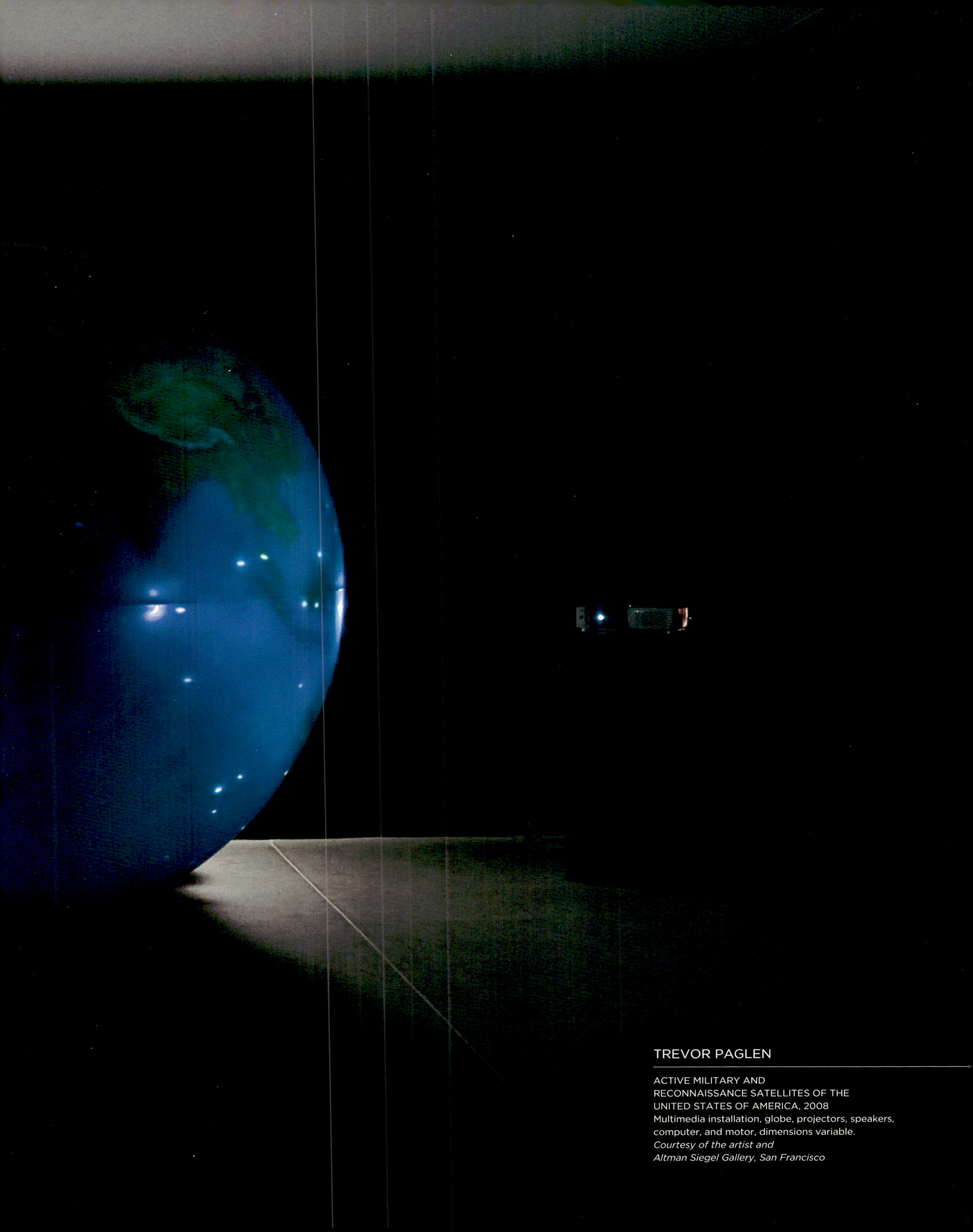

TREVOR PAGLEN

ACTIVE MILITARY AND
RECONNAISSANCE SATELLITES OF THE
UNITED STATES OF AMERICA, 2008
Multimedia installation, globe, projectors, speakers,
computer, and motor, dimensions variable.
Courtesy of the artist and
Altman Siegel Gallery, San Francisco

JED LIND

BETTER TO BURN OUT NO.1, 2008
C-print, 63 x 63 cm.
Private collection, courtesy
Jessica Bradley Art + Projects

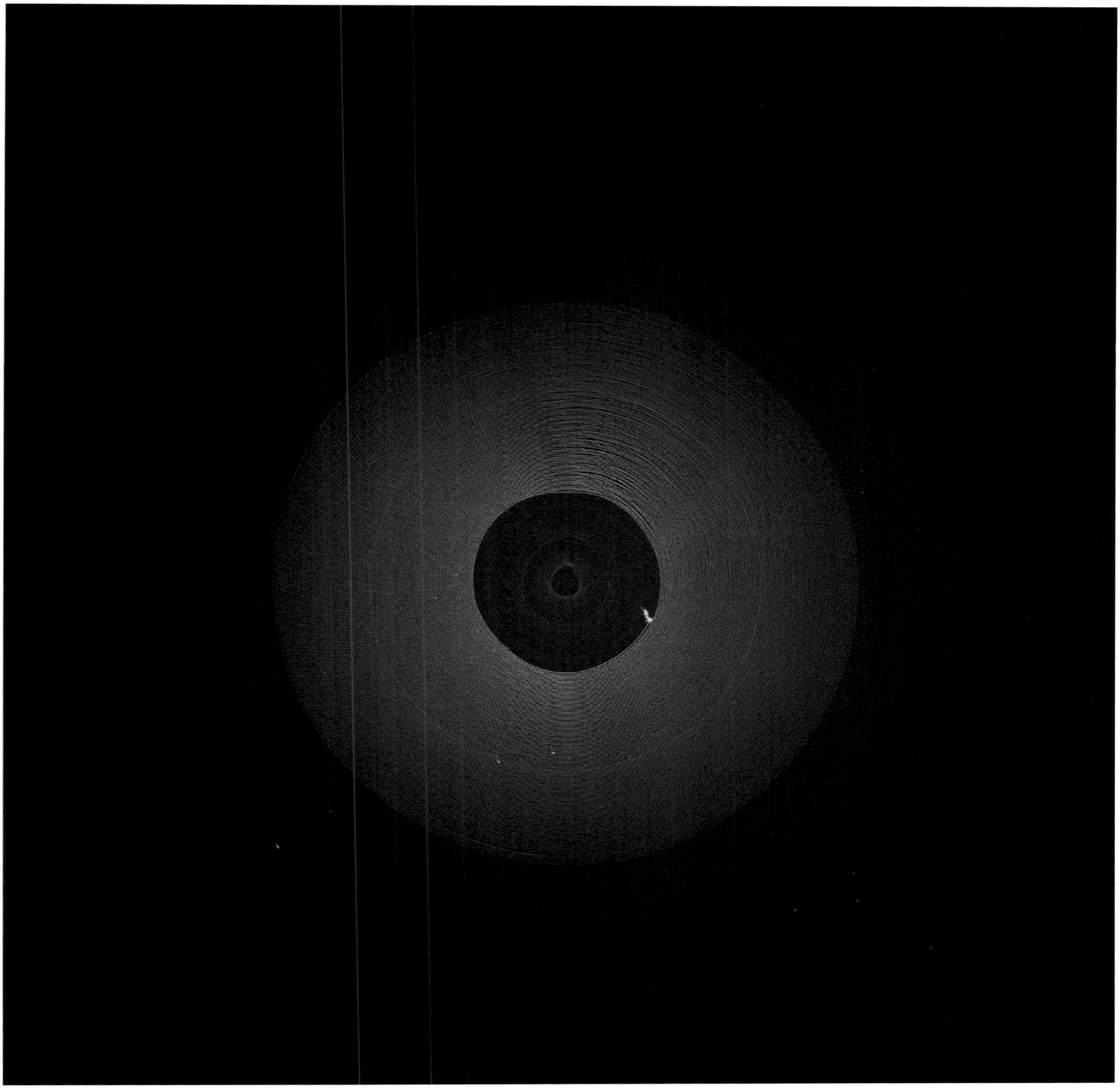

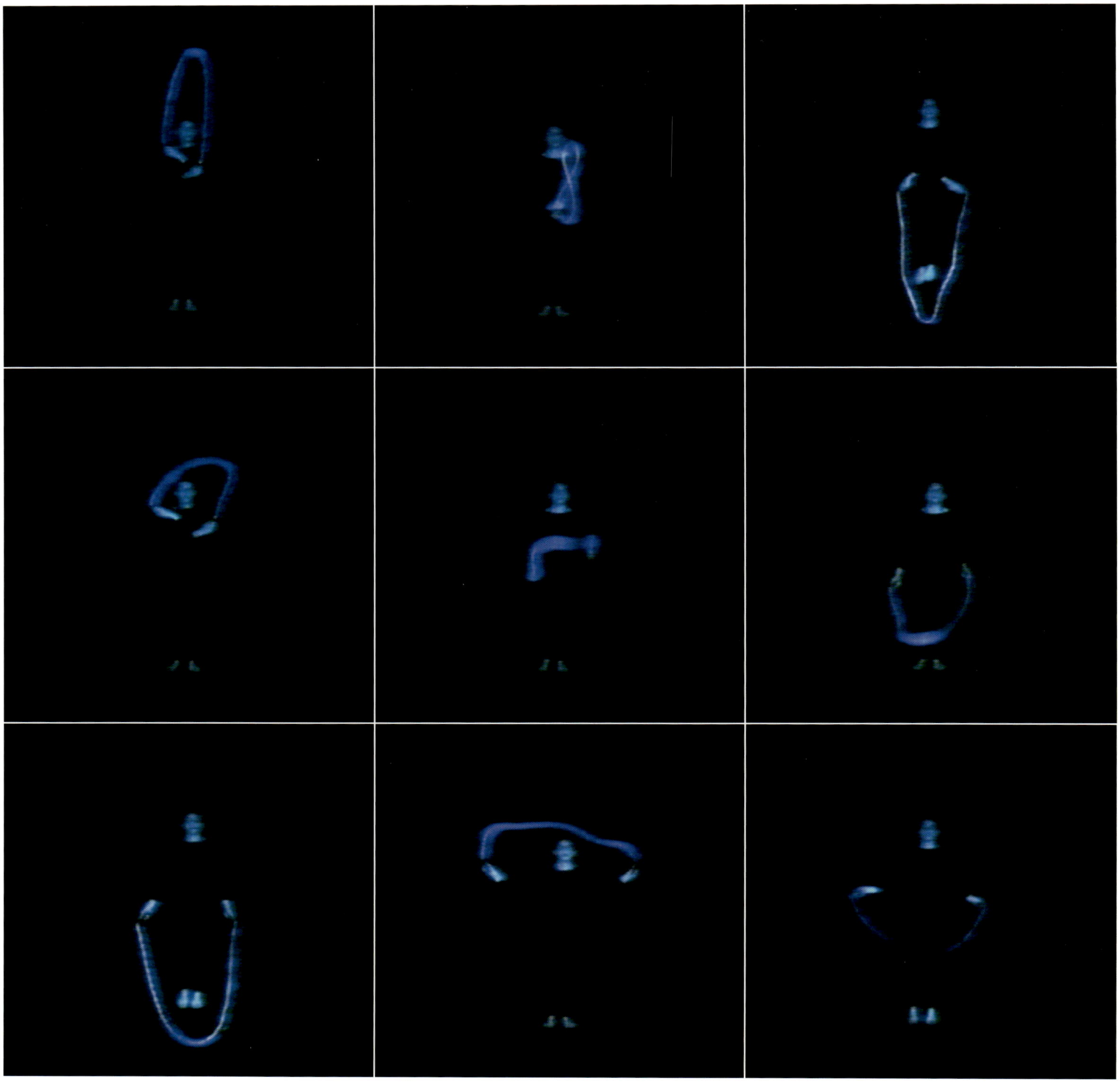

ANTONIA HIRSCH

STRING THEORY, 2003, stills (above)
Digital video and audio installation,
dimensions variable.
Collection of the artist

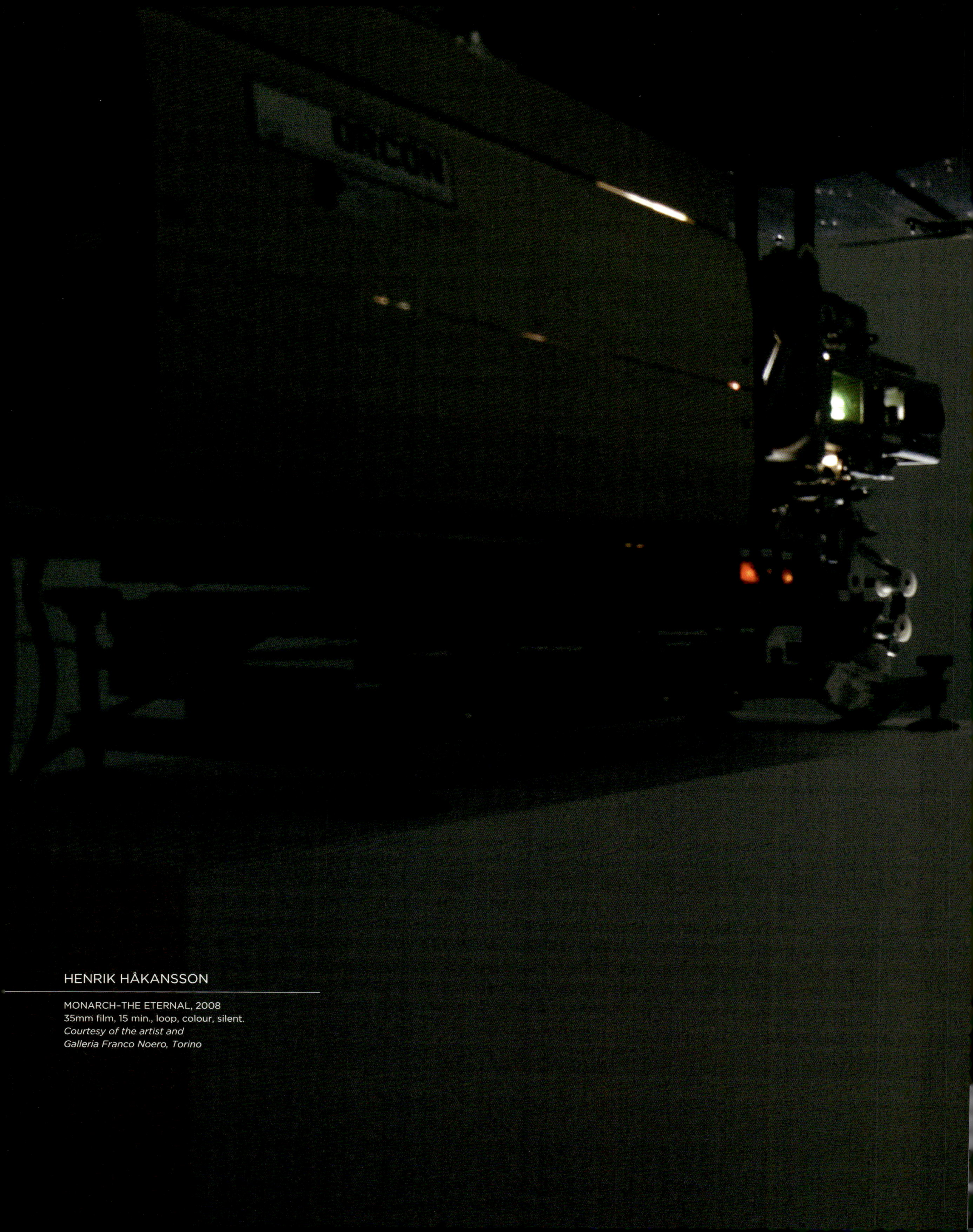

HENRIK HÅKANSSON

MONARCH–THE ETERNAL, 2008
35mm film, 15 min., loop, colour, silent.
Courtesy of the artist and
Galleria Franco Noero, Torino

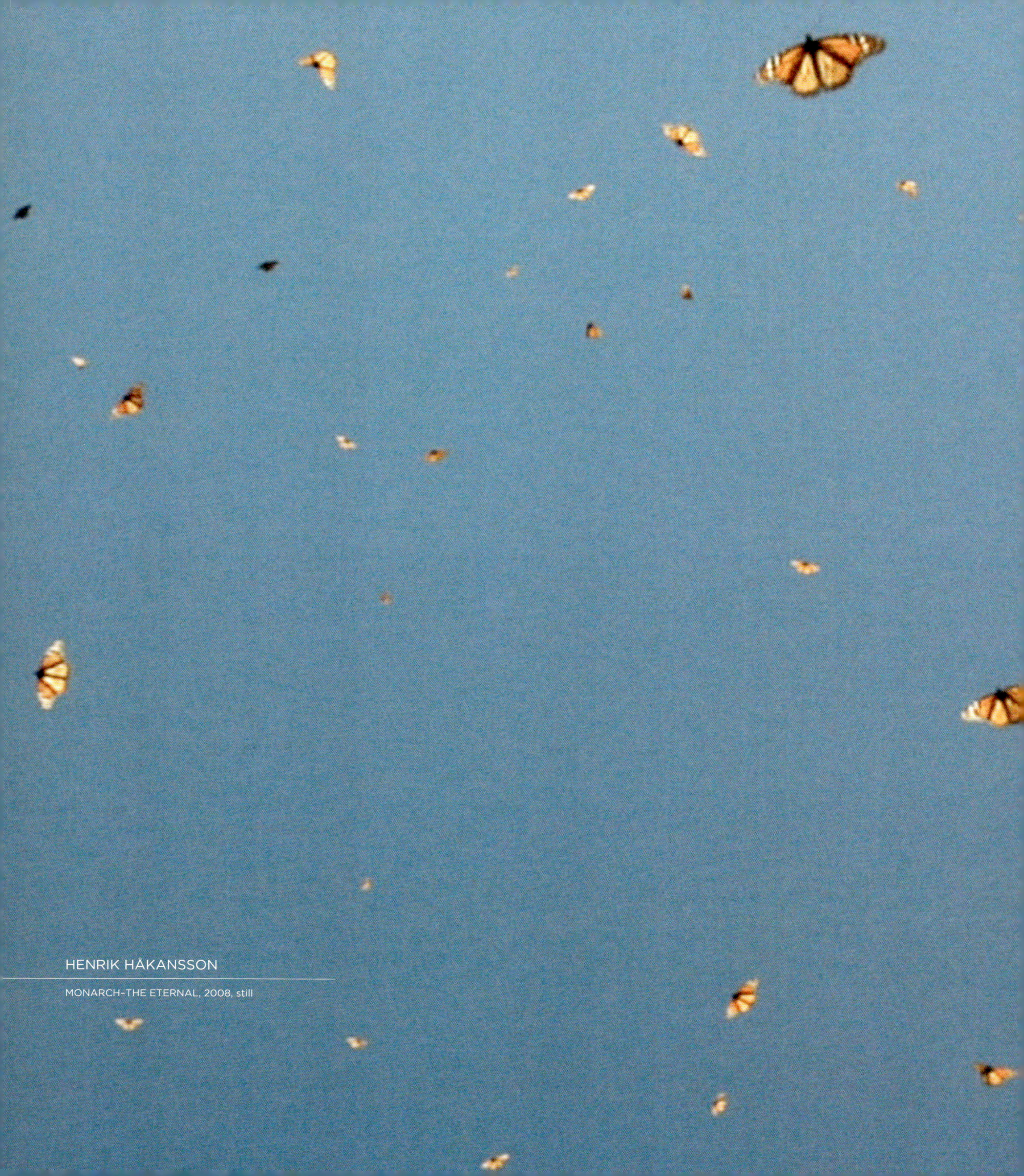

HENRIK HÅKANSSON

MONARCH–THE ETERNAL, 2008, still

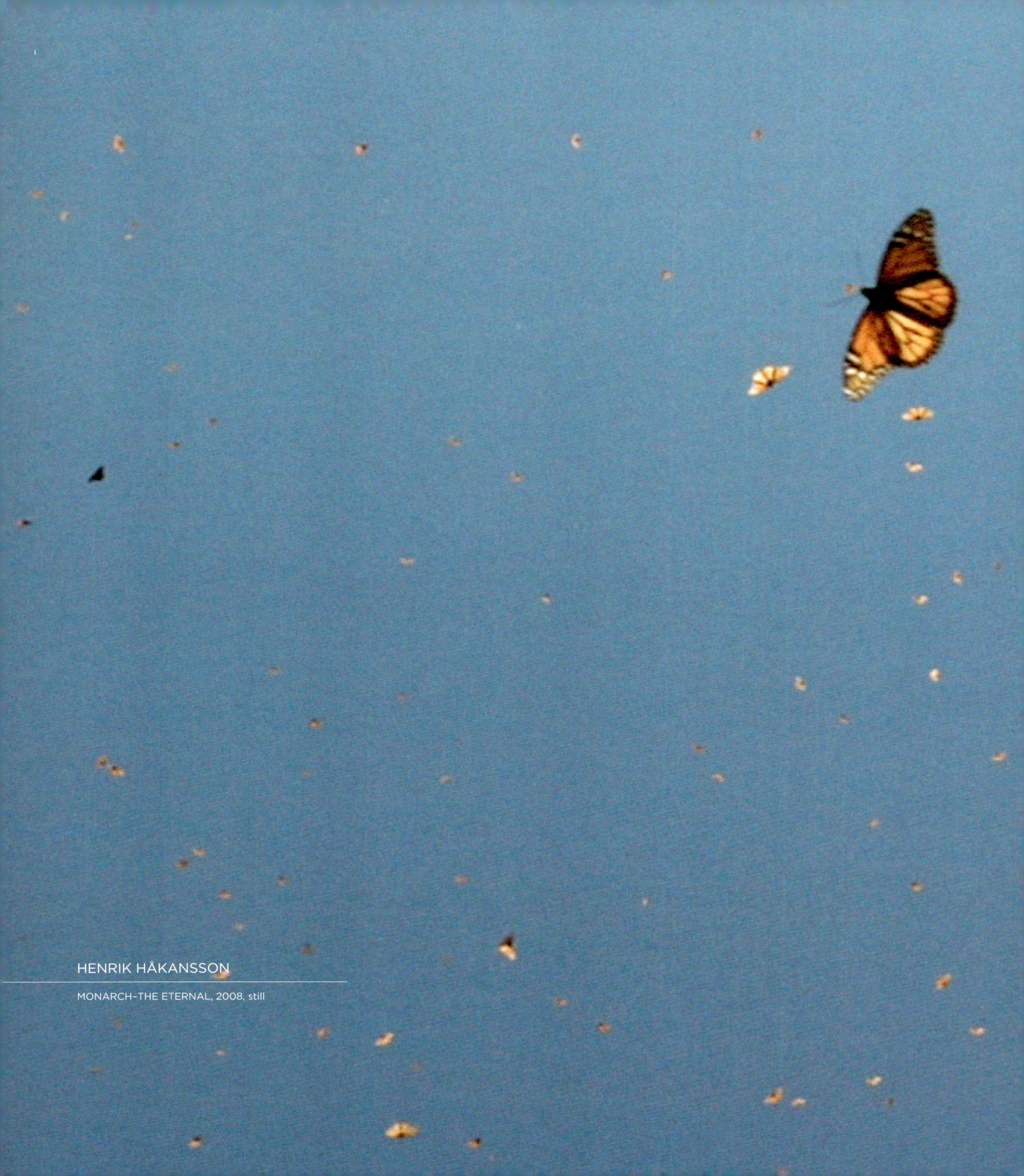

HENRIK HÅKANSSON

MONARCH–THE ETERNAL, 2008, still

GABRIEL OROZCO

BLACK KITES PERSPECTIVE
(FRONT HORIZONTAL), 1997
C-print, 40.6 x 50.8 cm.
Courtesy of the artist and
kurimanzutto, Mexico City.
Collection Thomas H. Bjarnason

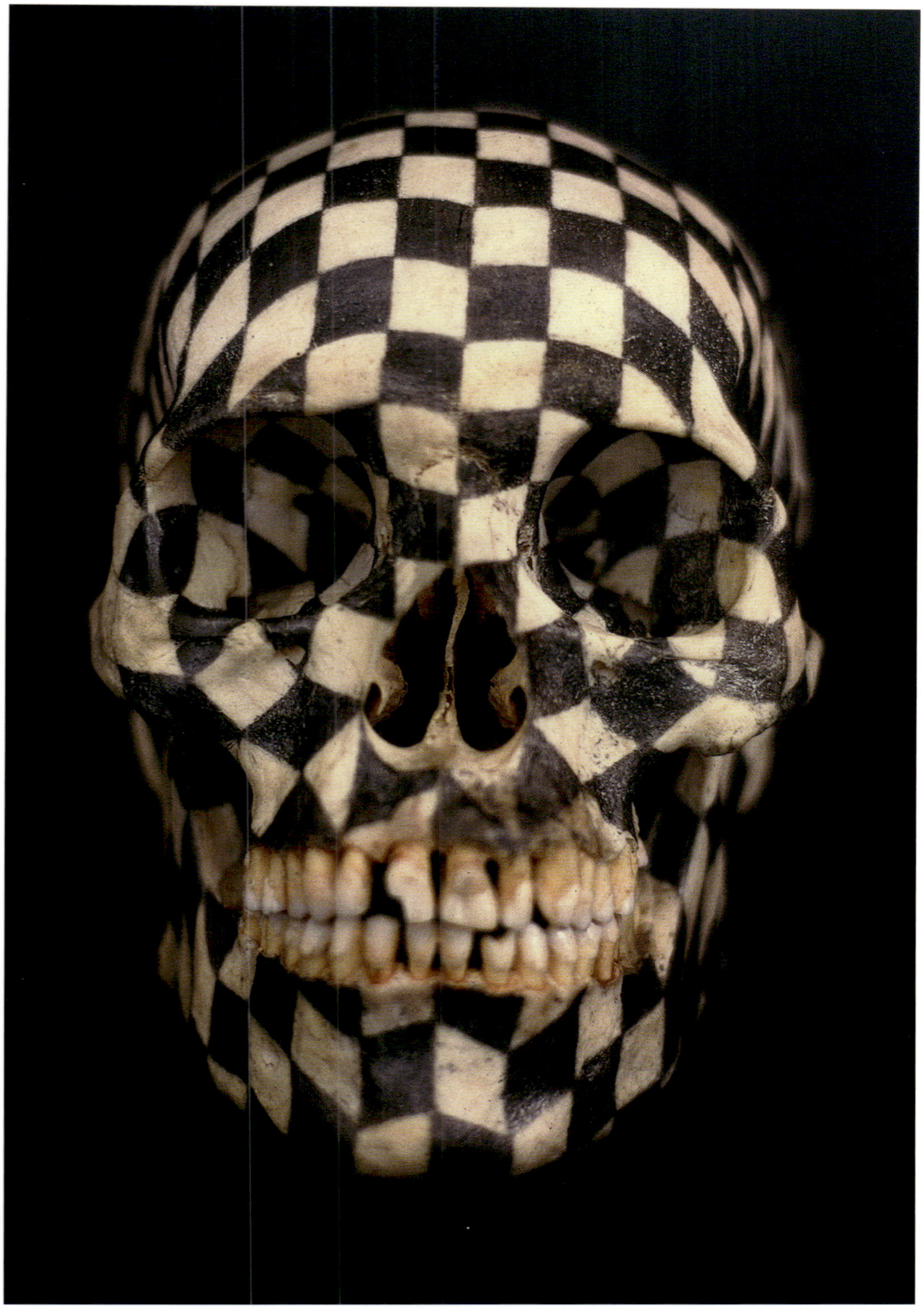

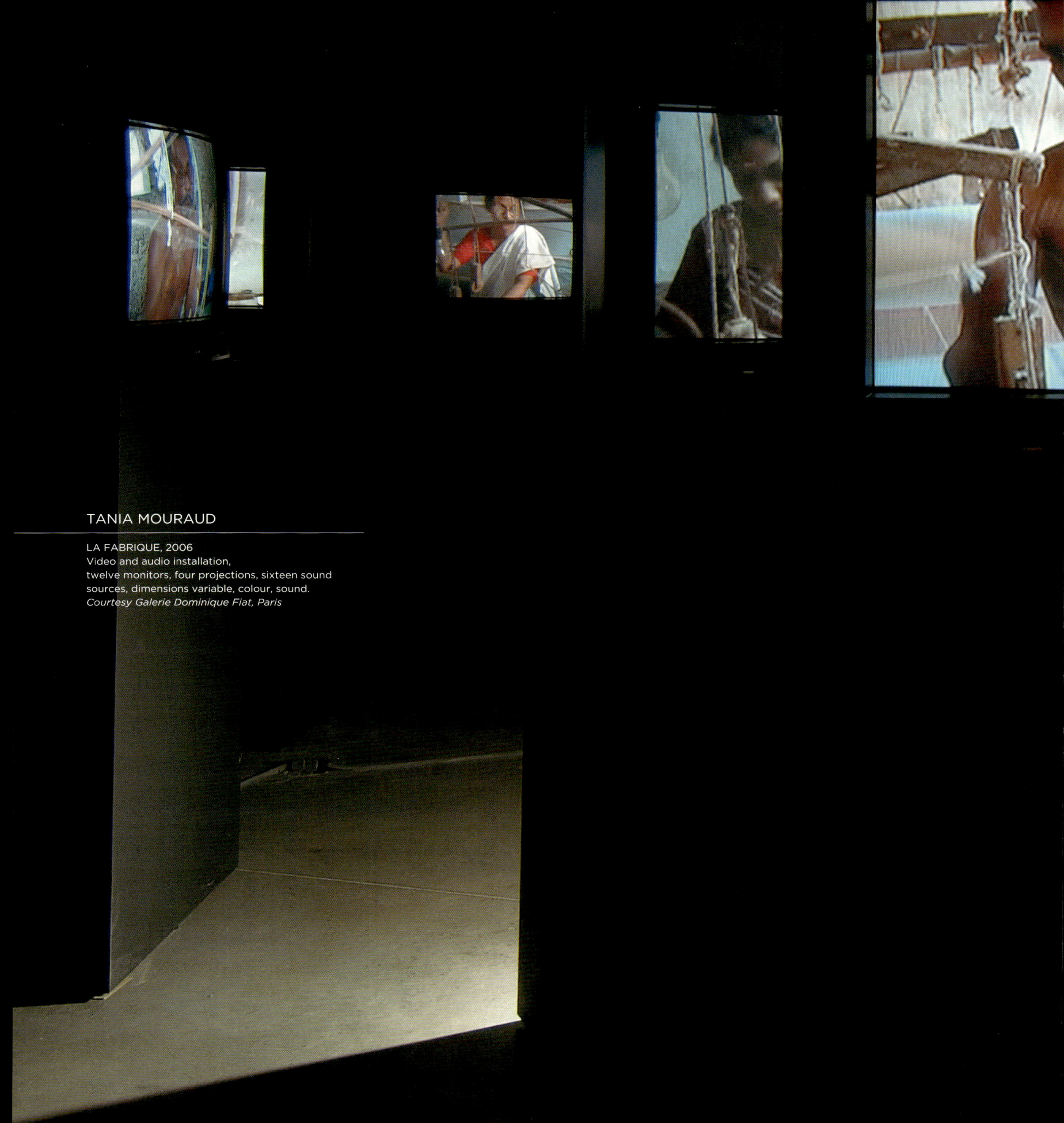

TANIA MOURAUD

LA FABRIQUE, 2006
Video and audio installation,
twelve monitors, four projections, sixteen sound sources, dimensions variable, colour, sound.
Courtesy Galerie Dominique Fiat, Paris

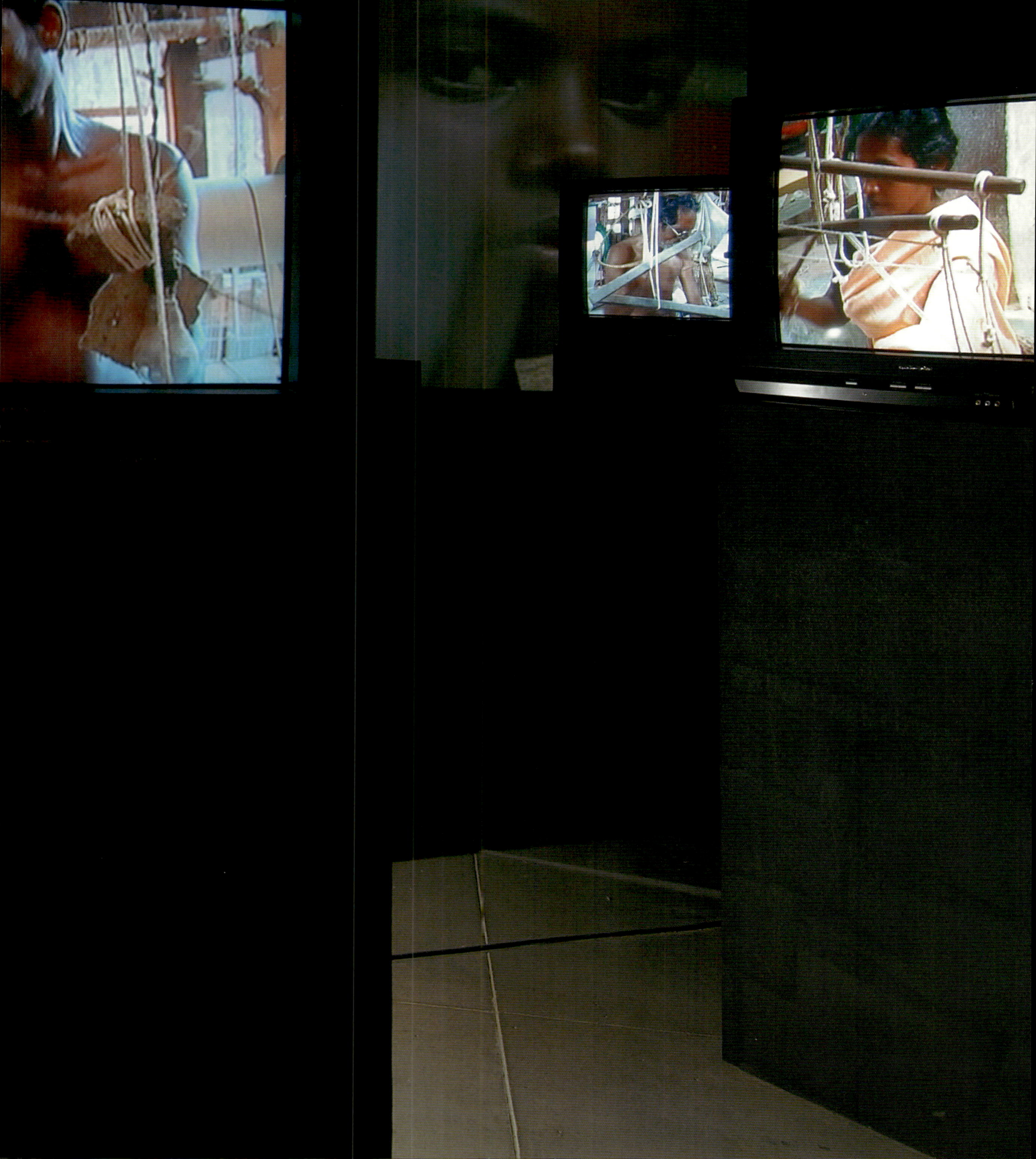

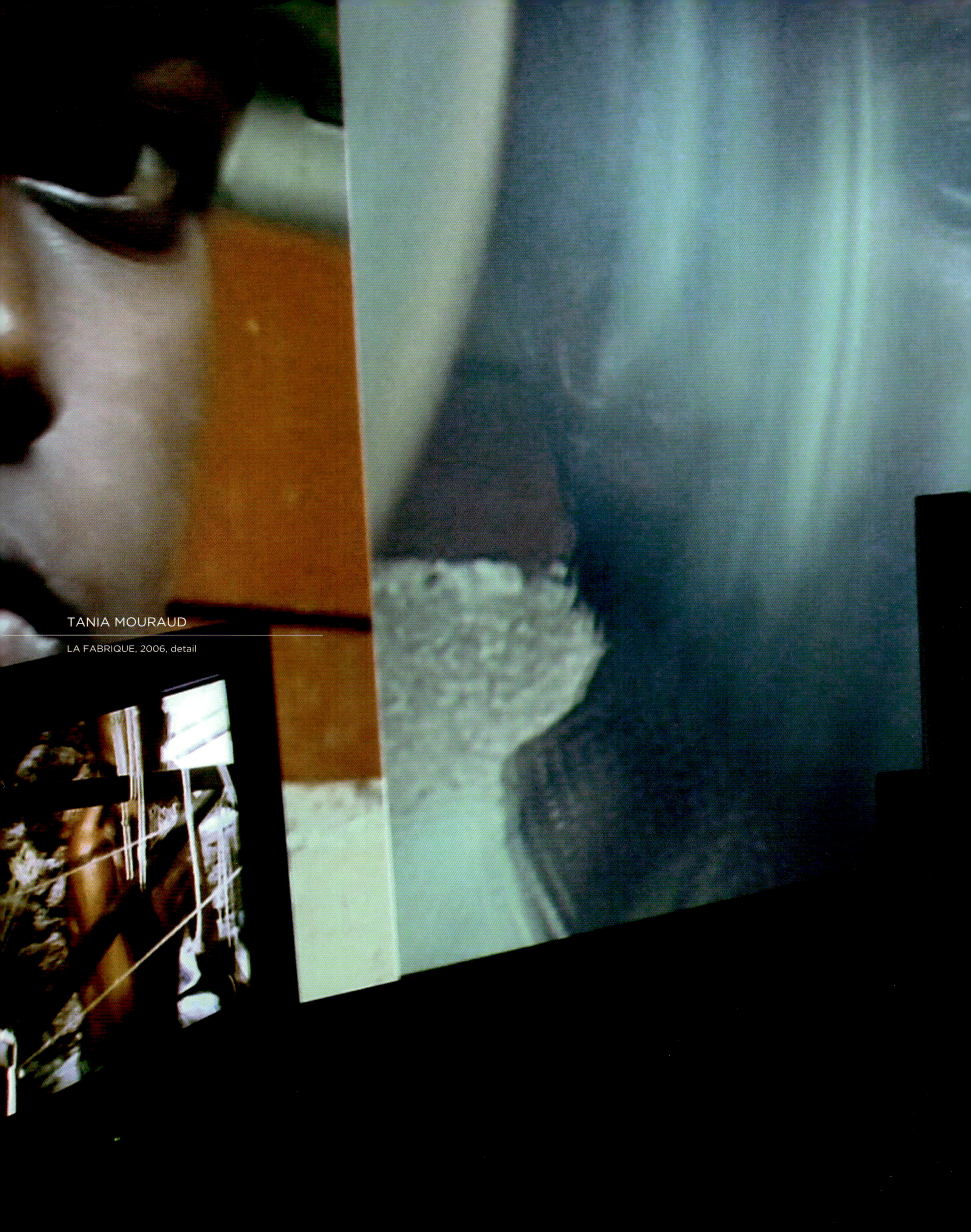

TANIA MOURAUD

LA FABRIQUE, 2006, detail

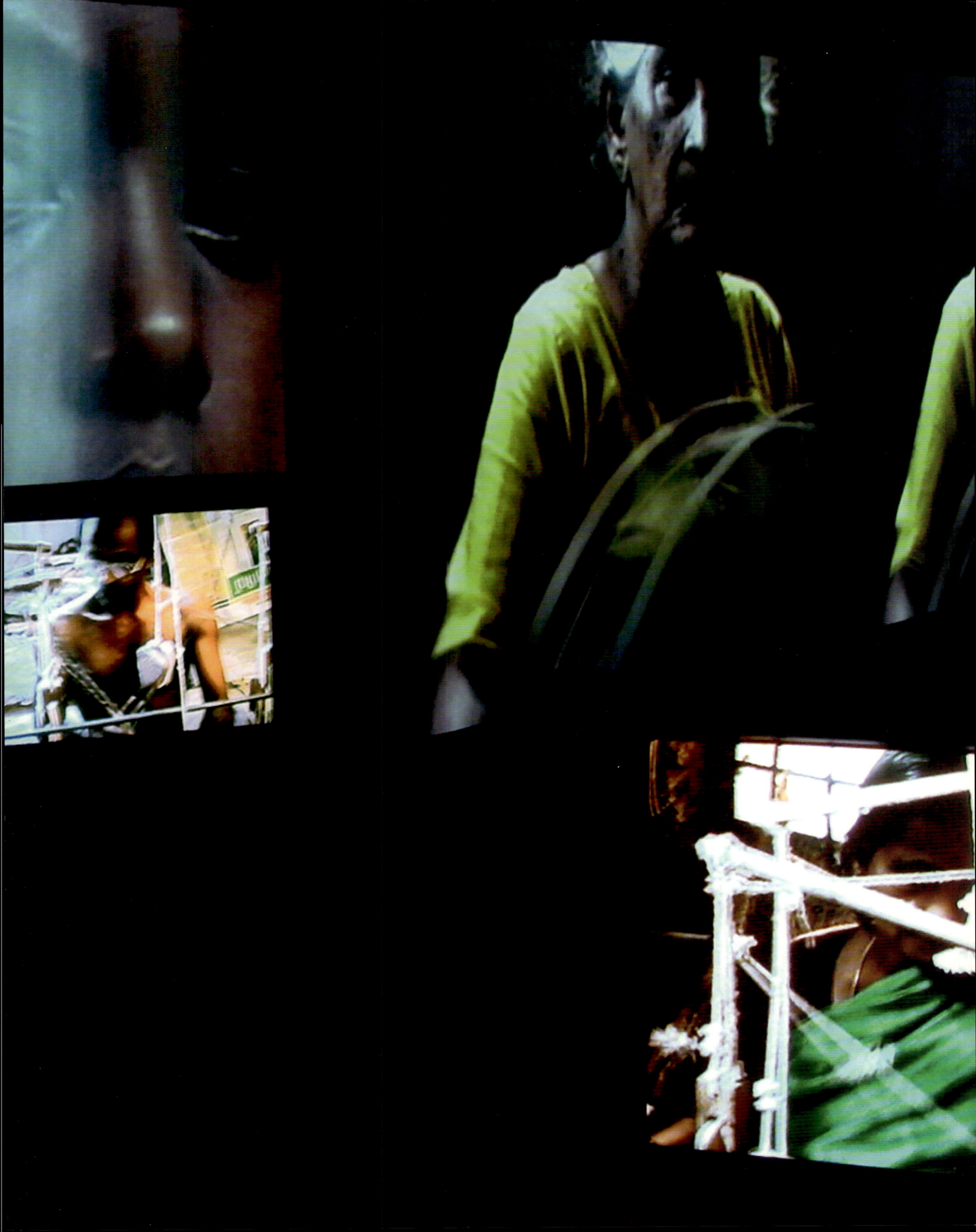

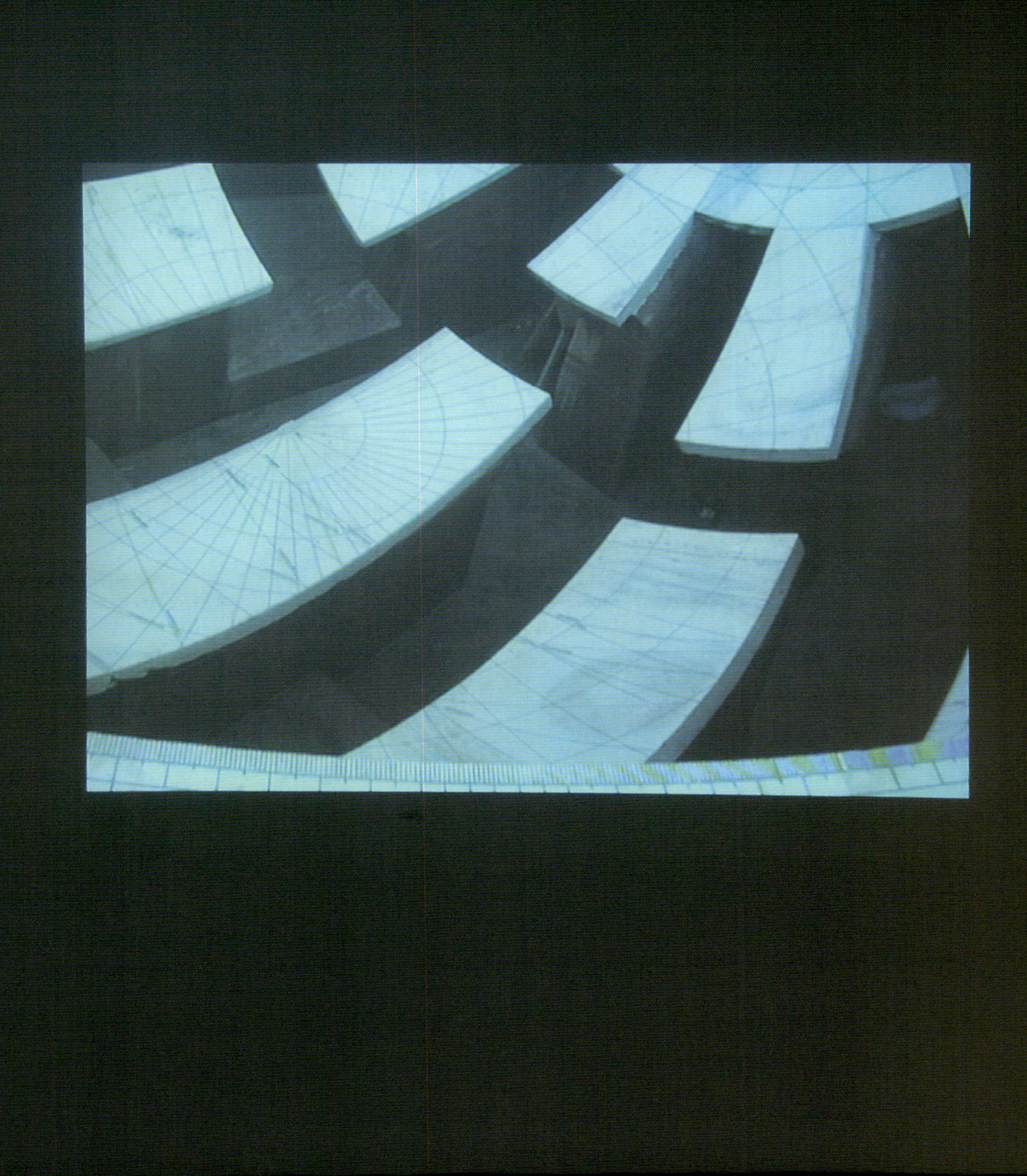

ANN VERONICA JANSSENS

JAIPRAKASH YANTRA, 2008, stills (right)
Video projection, 1 min., 28 sec., loop, colour, silent.
Courtesy Galerie Micheline Szwajcer, Antwerp

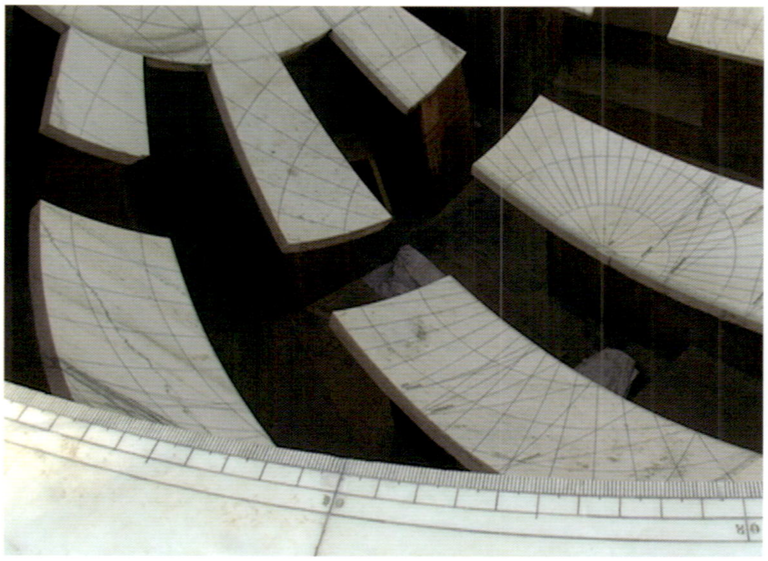
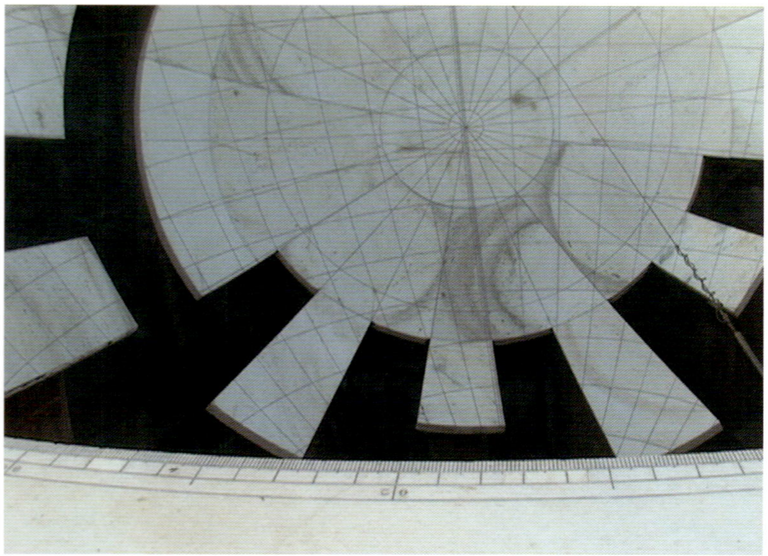
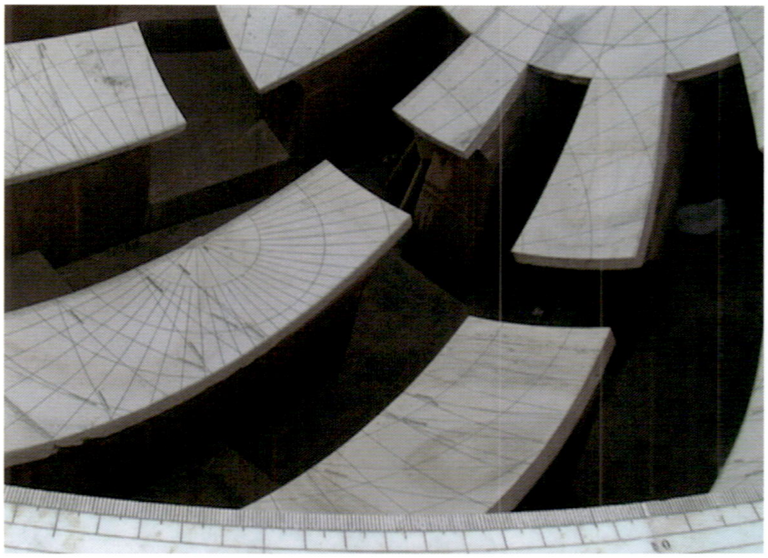
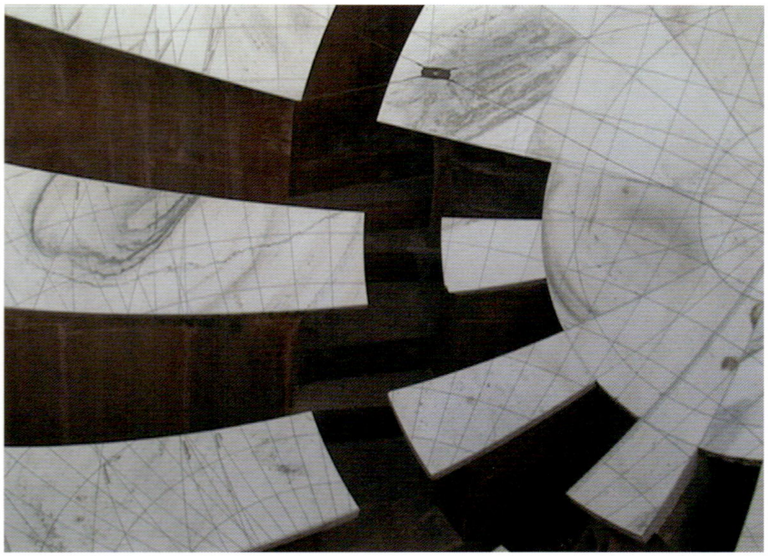
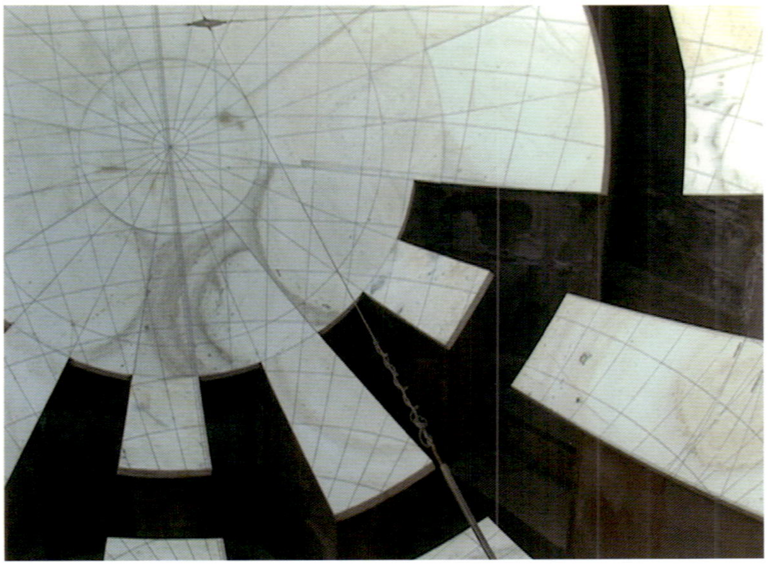
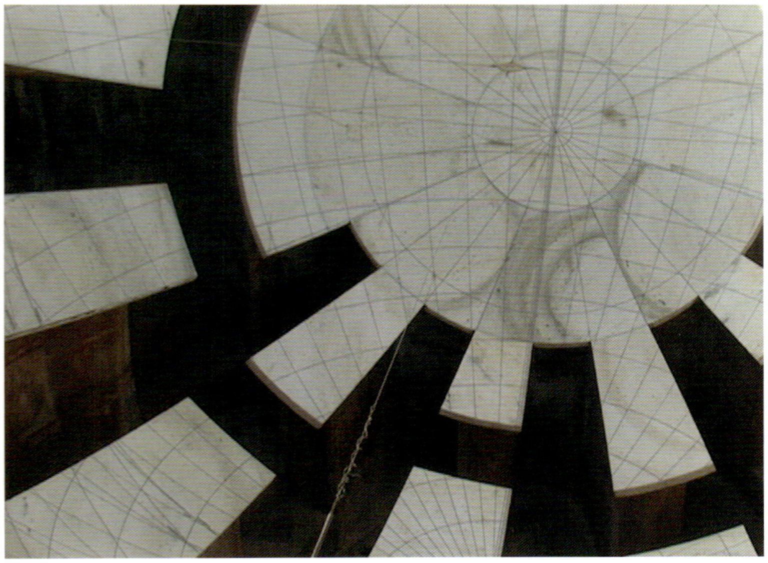

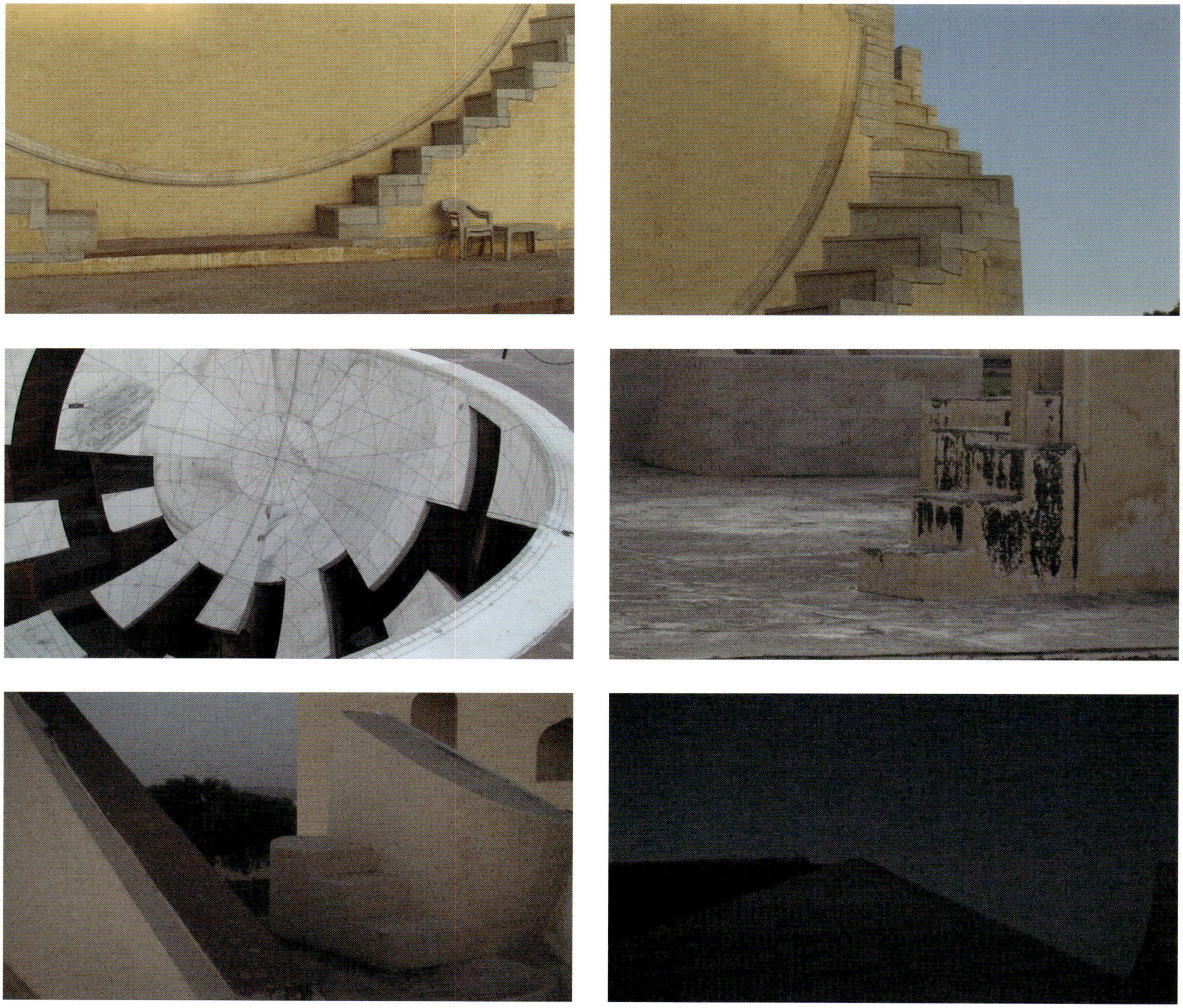

ANN VERONICA JANSSENS

CRÉPUSCULE À JANTAR MANTAR (VÉNUS),
2008, stills (above)
Single-monitor installation, 5 min., 13 sec.,
loop, colour, silent.
Courtesy Galerie Micheline Szwajcer, Antwerp

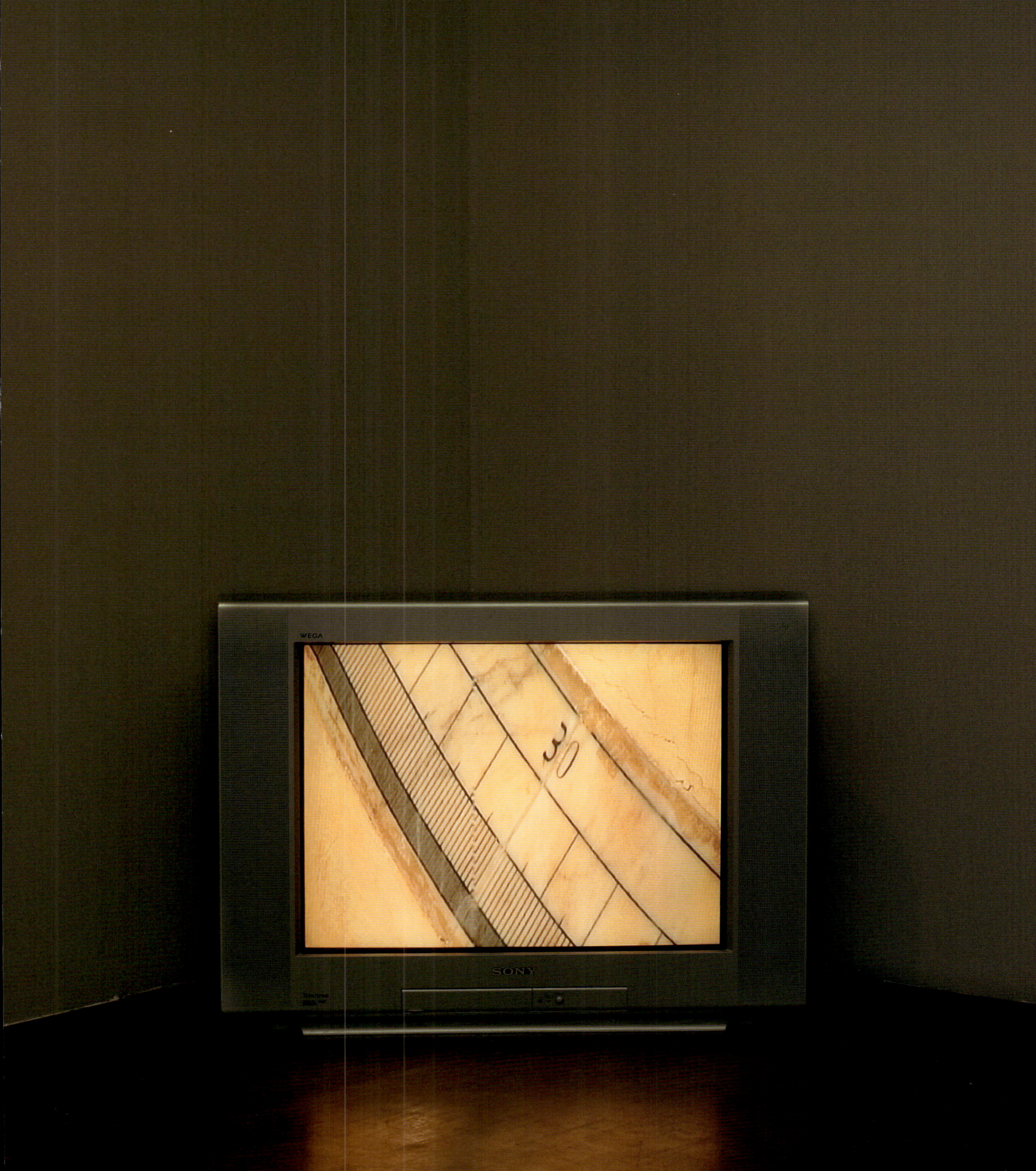
WEGA
SONY

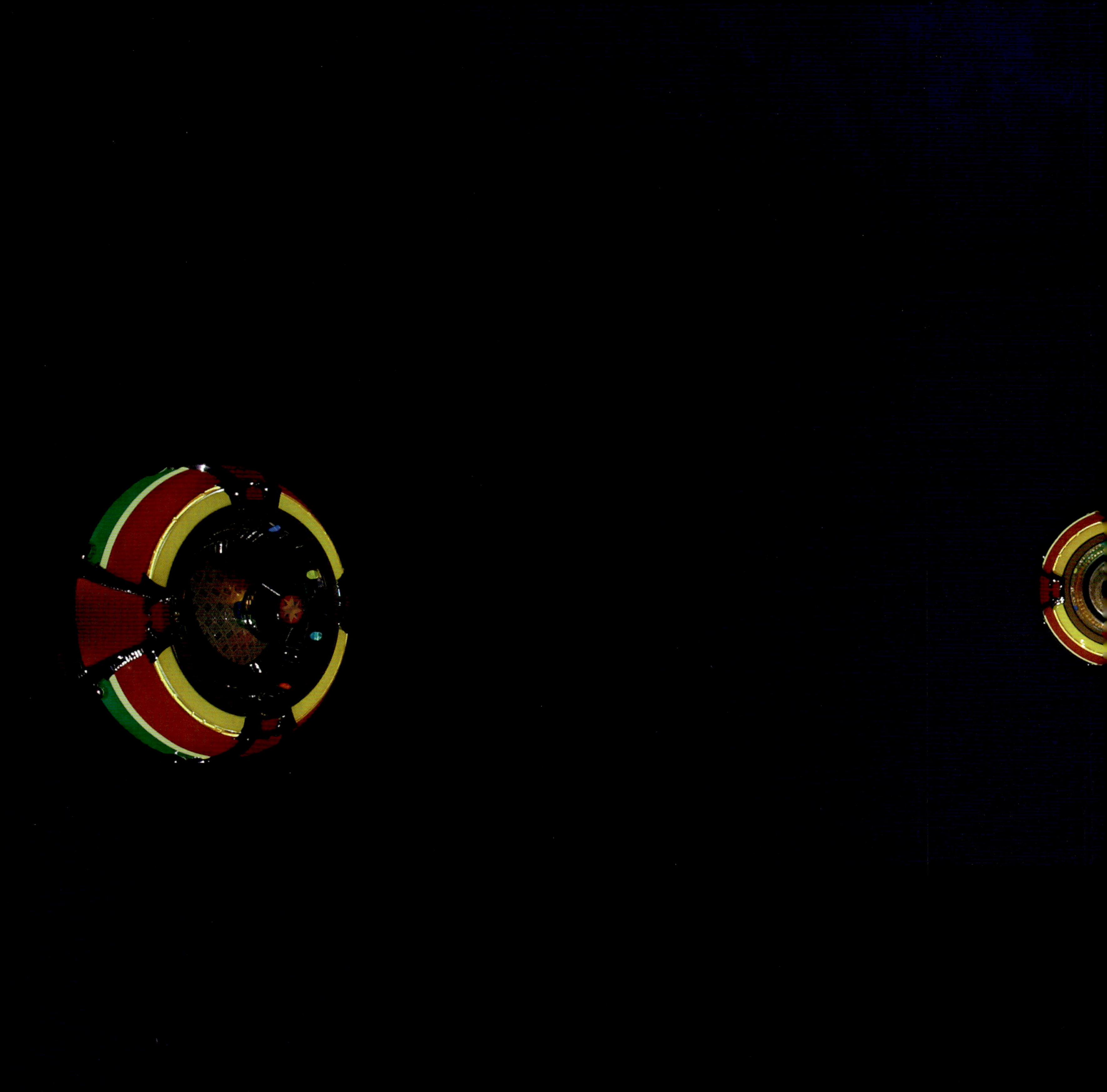

KIMSOOJA

MANDALA–ZONE OF ZERO, 2003
Installation of three jukeboxes, three-channel sound, Tibetan chant, Gregorian chant and Islamic chant.
Courtesy of the artist

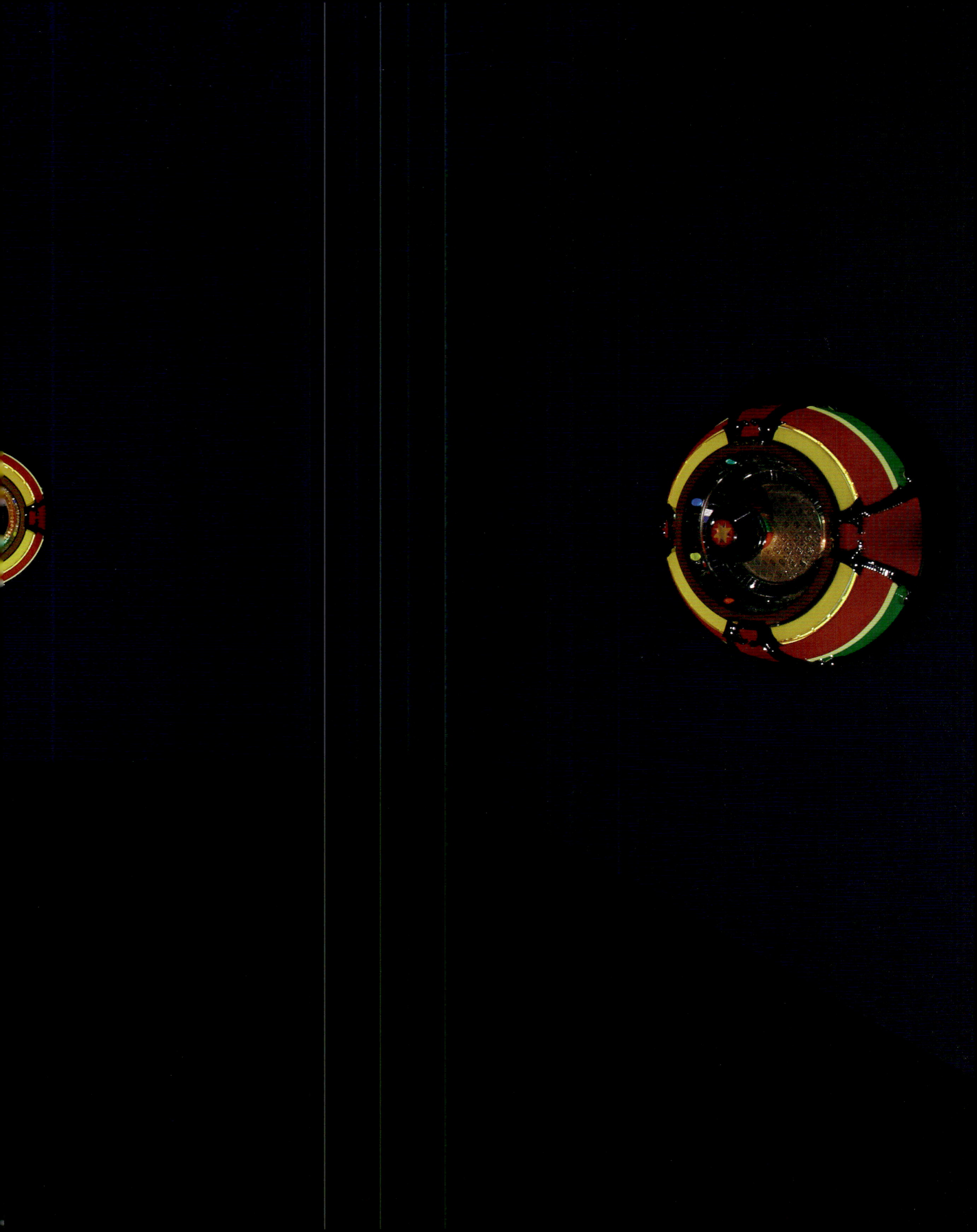

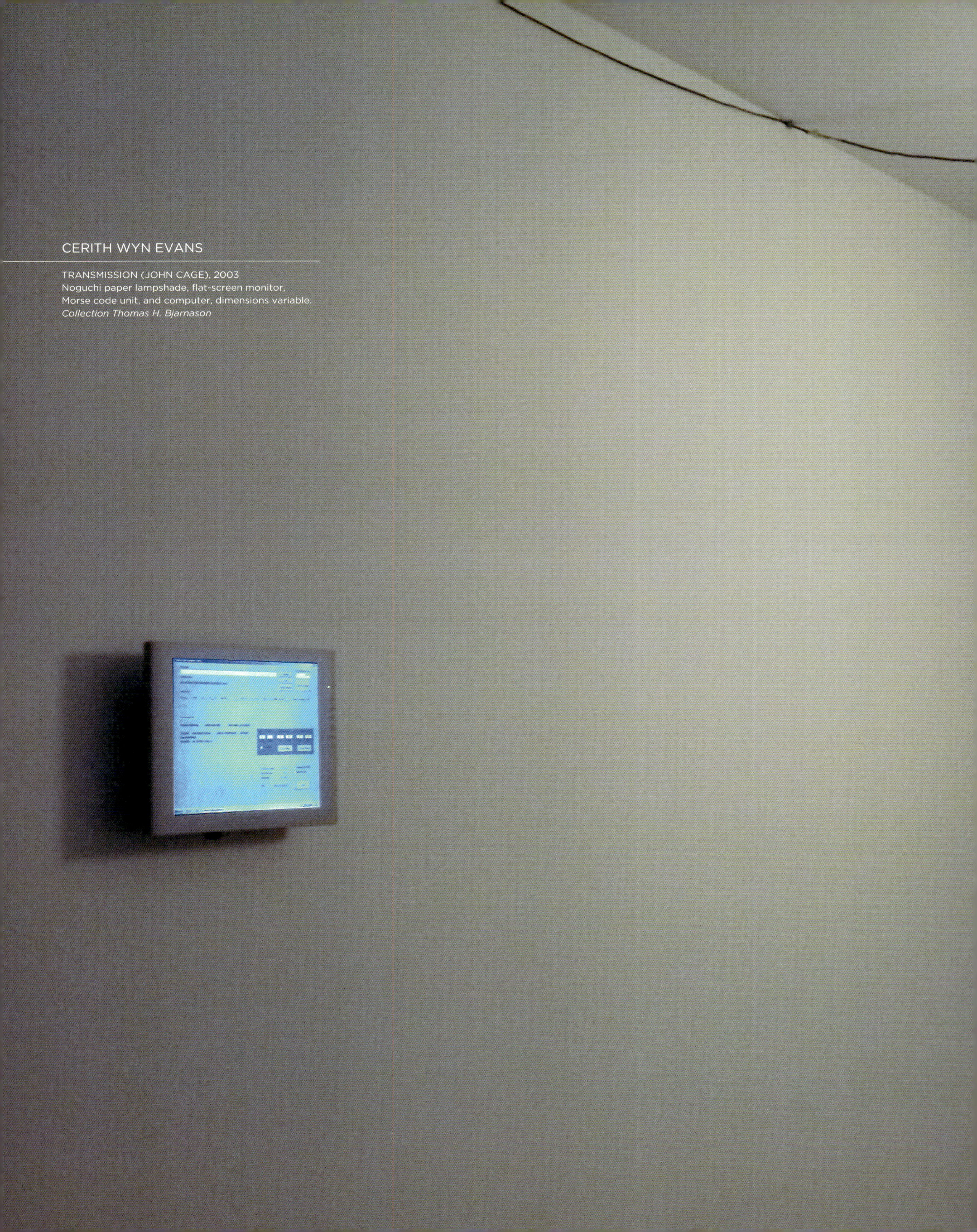

CERITH WYN EVANS

TRANSMISSION (JOHN CAGE), 2003
Noguchi paper lampshade, flat-screen monitor,
Morse code unit, and computer, dimensions variable.
Collection Thomas H. Bjarnason

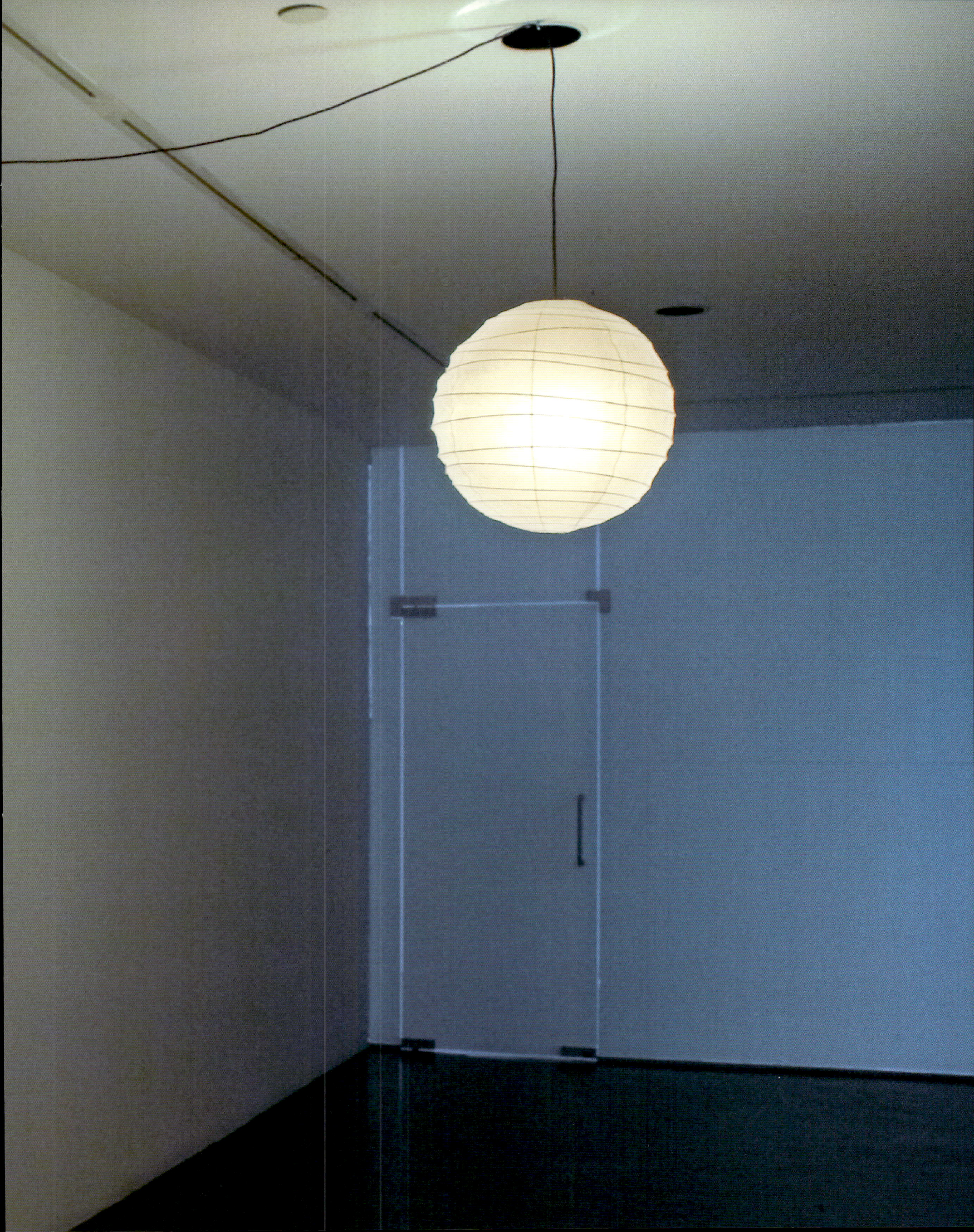

CERITH WYN EVANS

TRANSMISSION (JOHN CAGE), 2003, details

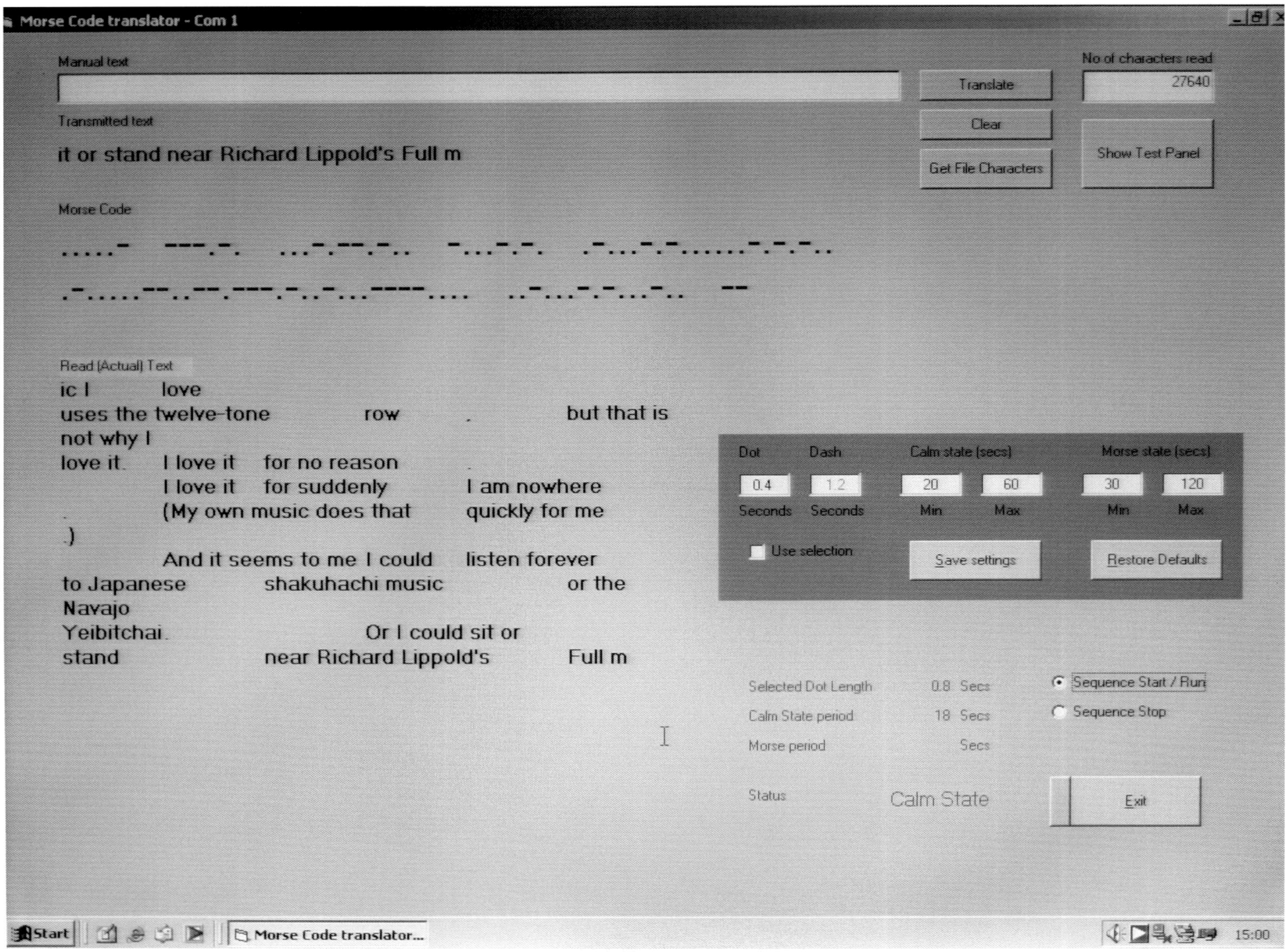

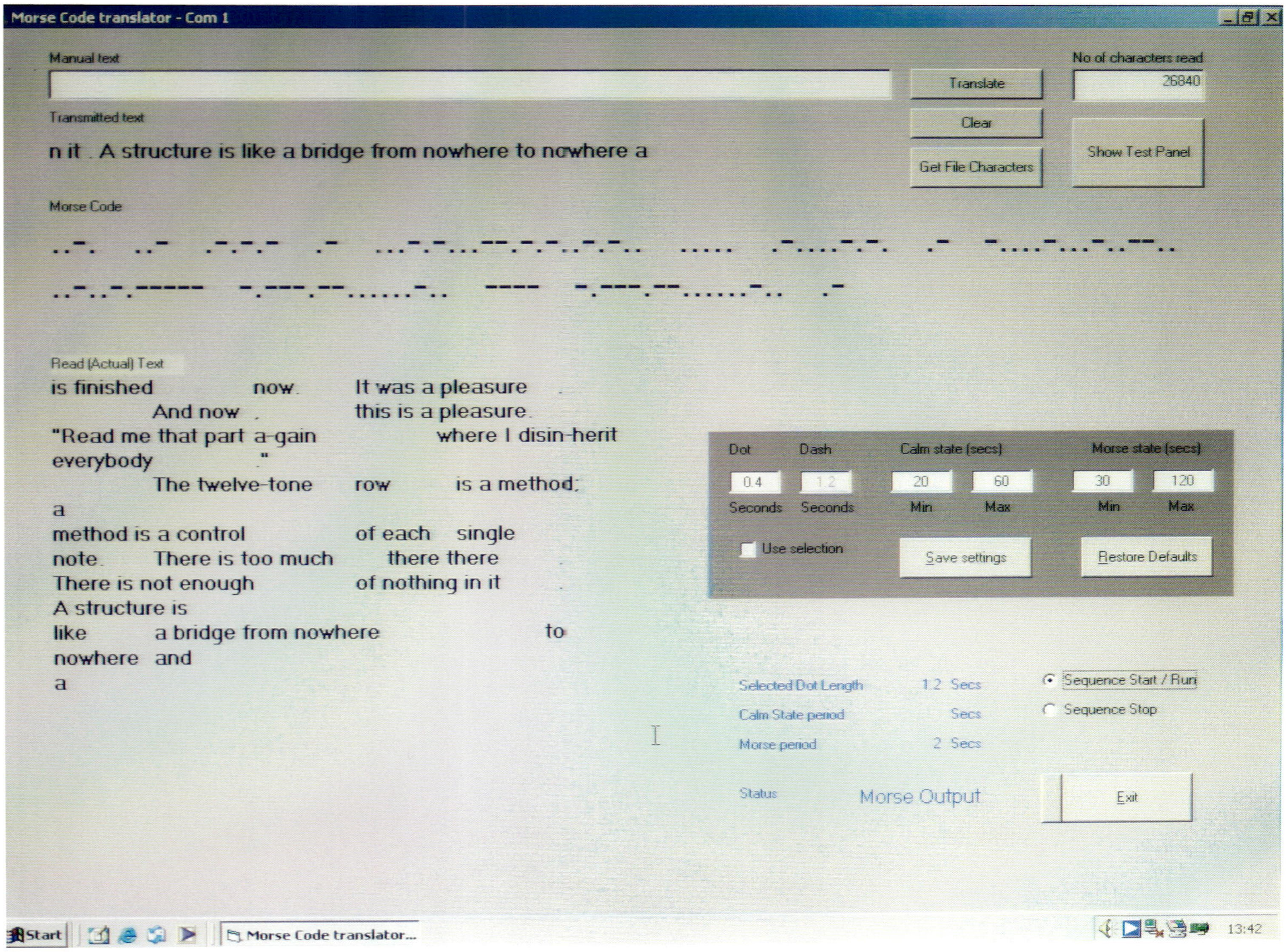

Morse Code translator - Com 1
Manual text
Translate
No of characters read
26840
Transmitted text
Clear
n it . A structure is like a bridge from nowhere to nowhere a
Get File Characters
Show Test Panel
Morse Code
Read (Actual) Text
is finished now. It was a pleasure
And now . this is a pleasure.
"Read me that part a-gain where I disin-herit
everybody ."
The twelve-tone row is a method;
a
method is a control of each single
note. There is too much there there
There is not enough of nothing in it
A structure is
like a bridge from nowhere to
nowhere and
a
Dot
Dash
Calm state (secs)
Morse state (secs)
0.4
1.2
20
60
30
120
Seconds
Seconds
Min
Max
Min
Max
Use selection
Save settings
Restore Defaults
Selected Dot Length
1.2 Secs
Sequence Start / Run
Calm State period
Secs
Sequence Stop
Morse period
2 Secs
Status
Morse Output
Exit
Start
Morse Code translator...
13:42

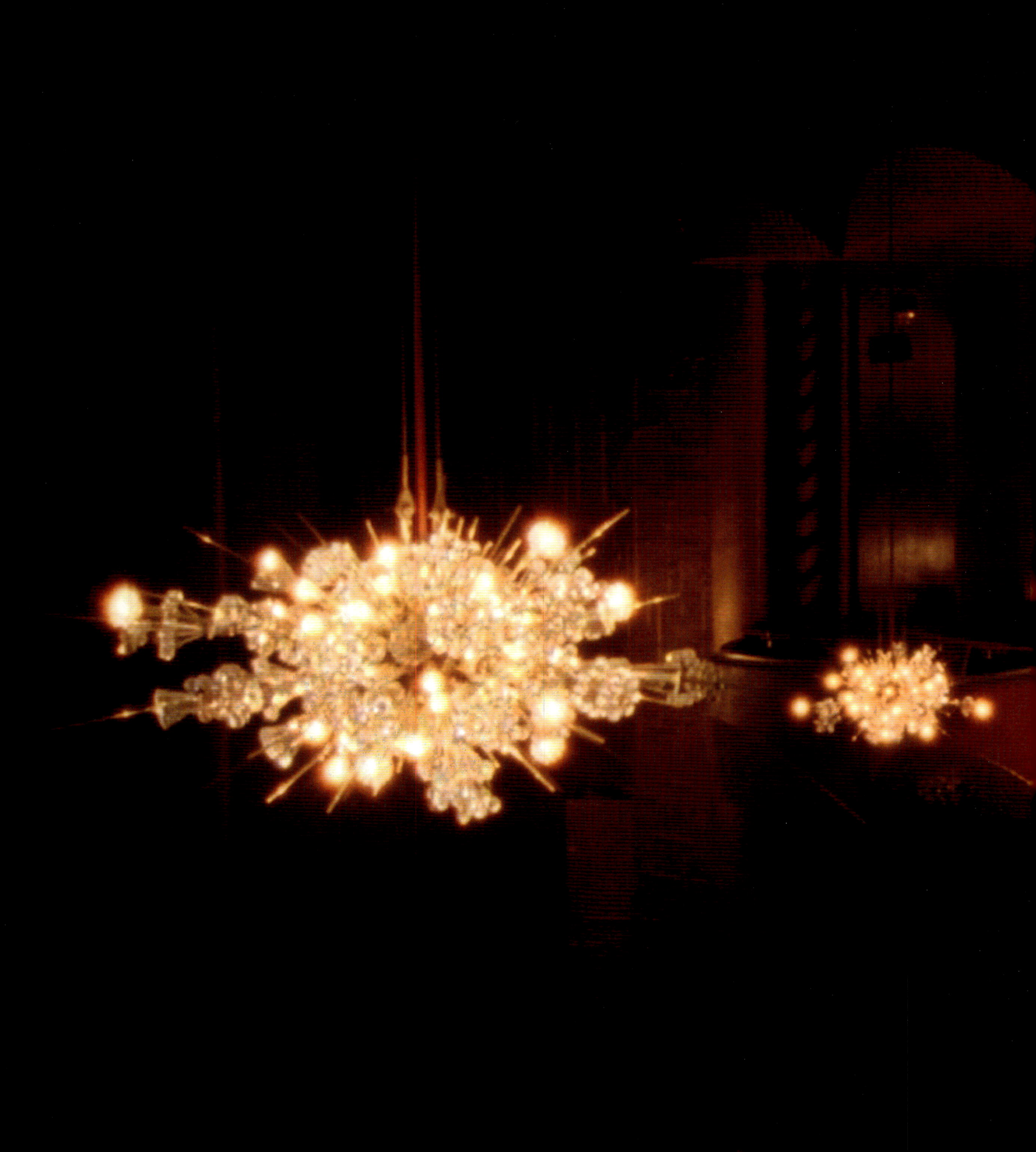

JOSIAH MCELHENY

ISLAND UNIVERSE, 2008, still
High-definition video projection of
super 16mm film, 9 min., colour, sound.
Courtesy White Cube, London

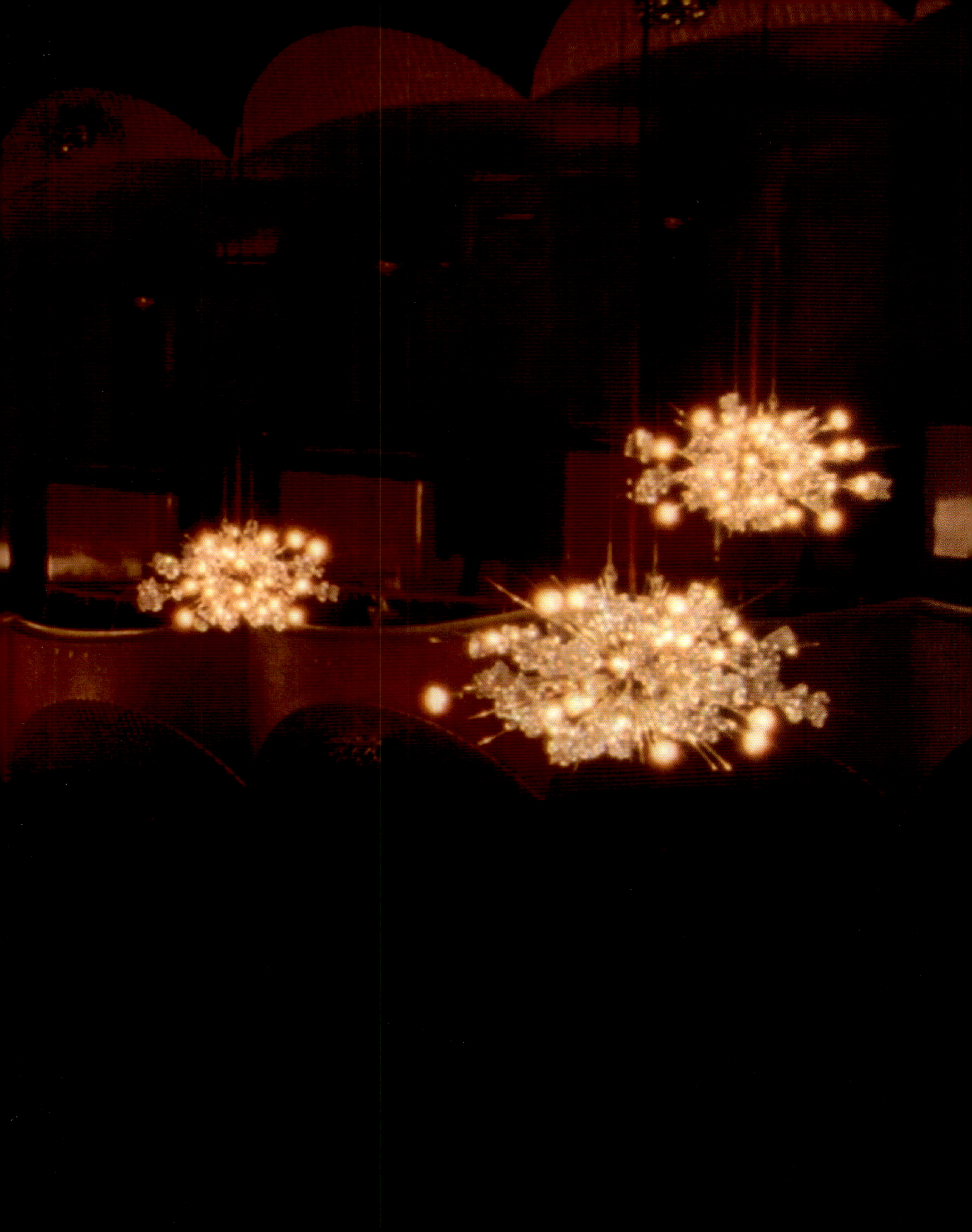

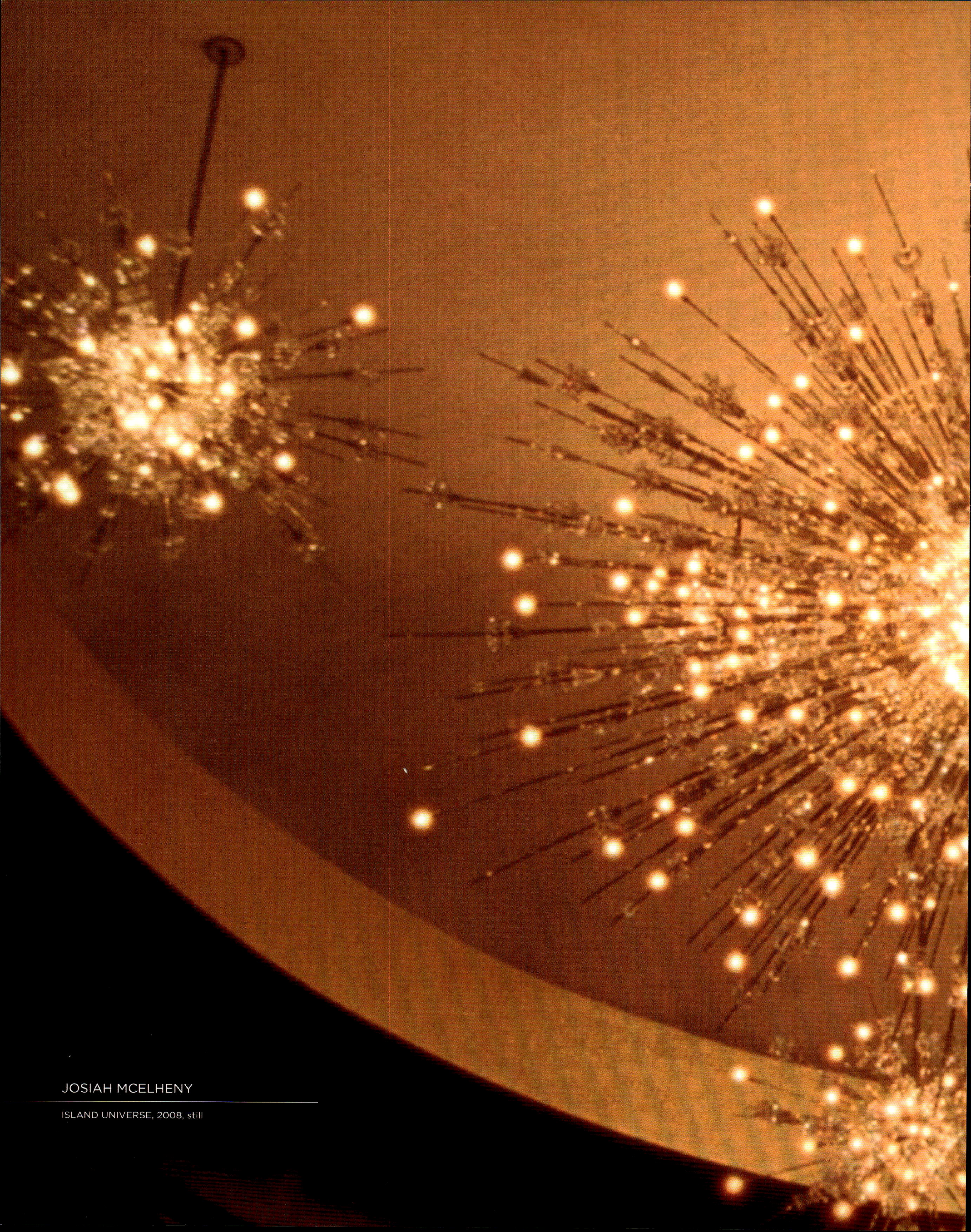

JOSIAH MCELHENY

ISLAND UNIVERSE, 2008, still

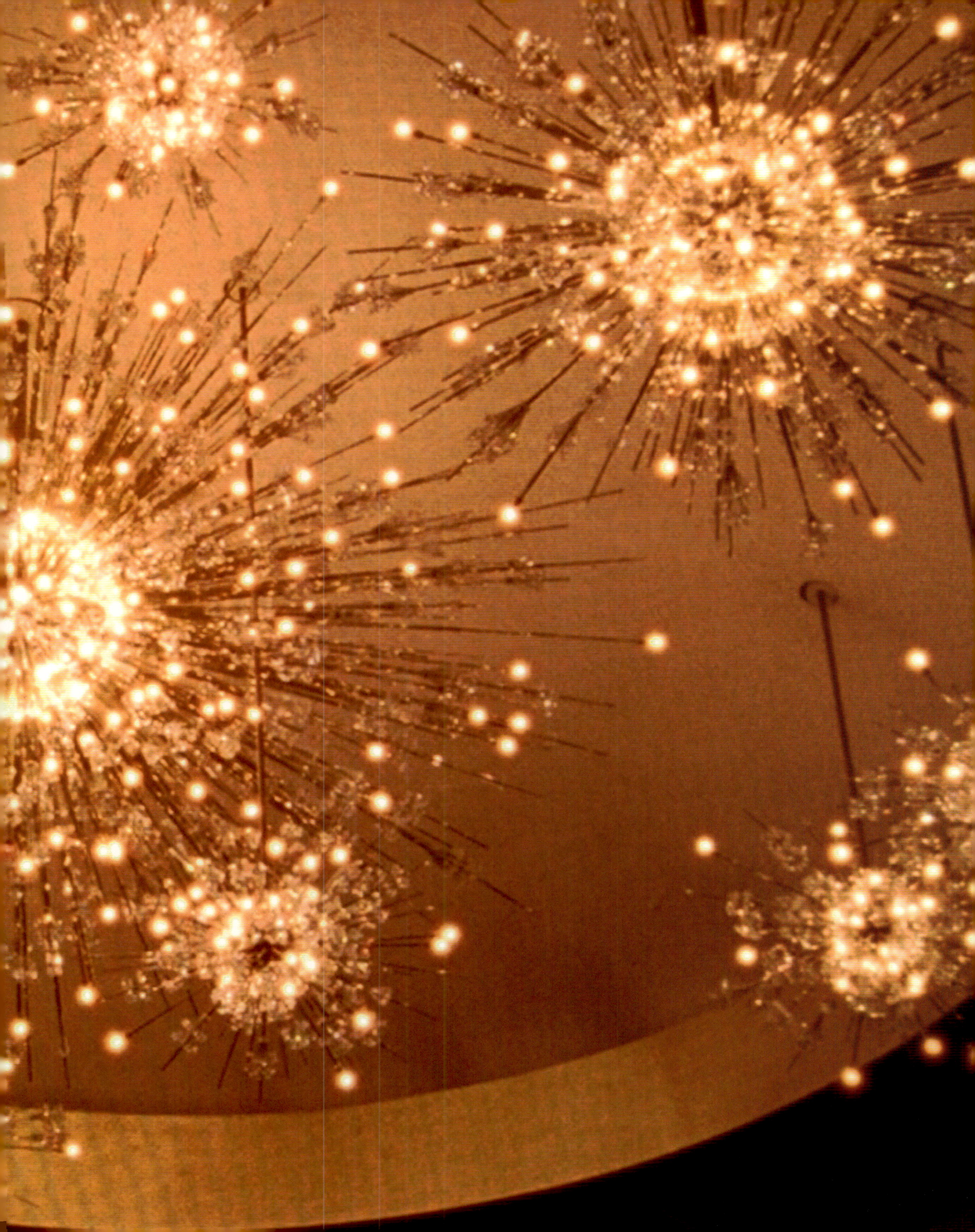

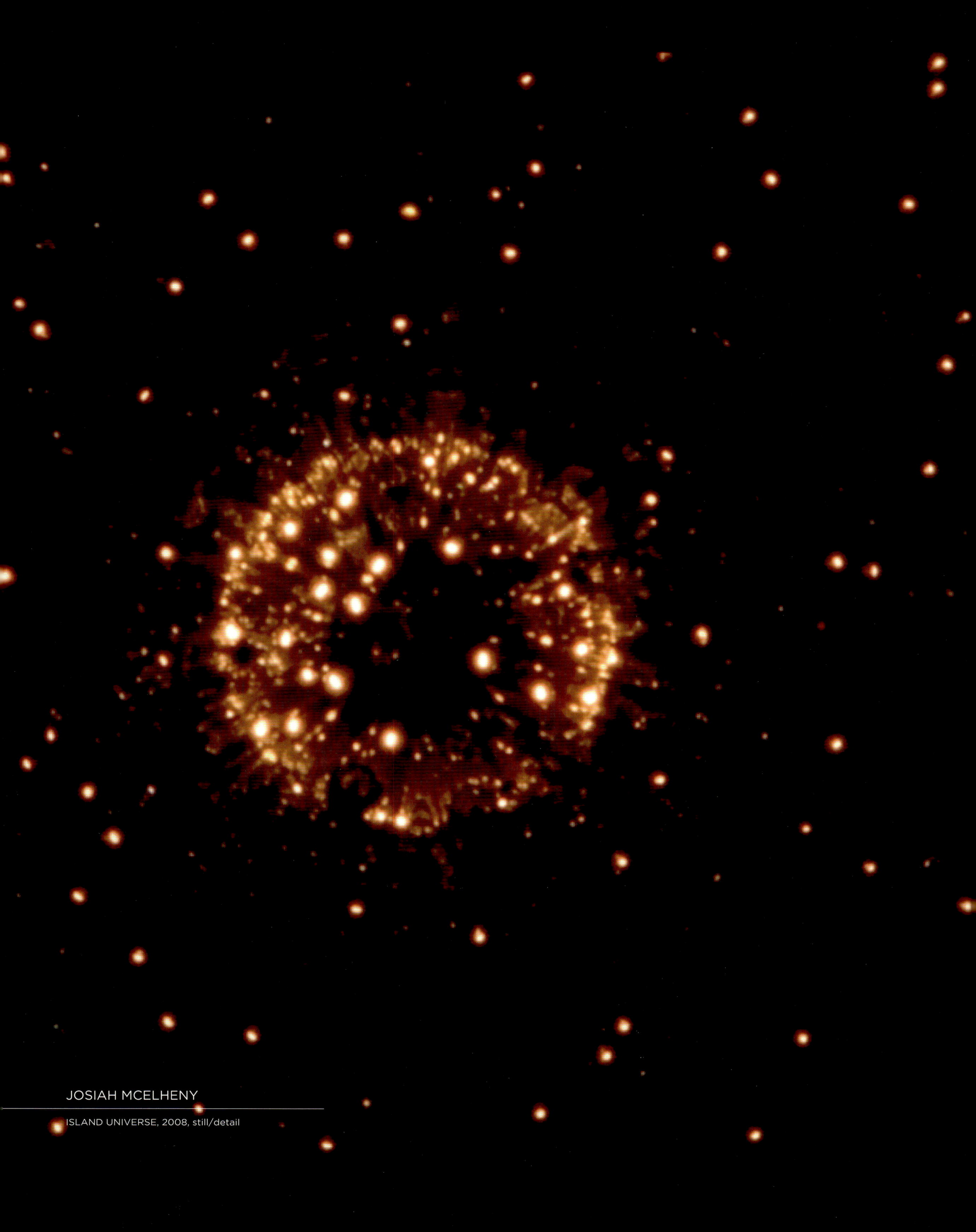

JOSIAH MCELHENY

ISLAND UNIVERSE, 2008, still/detail

THE ARTISTS

ADEL ABDESSEMED

Born in Constantine, Algeria, 1971. Lives in New York, USA.

Adel Abdessemed uses everyday materials and a wide range of media to create explosive and candid critiques of radical, political and religious ideologies. *God is Design*, 2005, is an animation composed of more than three thousand drawings, abstract representations that recall Western geometric abstraction and Eastern arabesques. As the diagrams mutate and morph into one another they come to represent, at different moments, the ornamental motifs of the monotheist religions Islam, Judaism and Christianity. The video's soundtrack, a collision of sound samples, was composed by Silvia Ocougne.

Produced in the aftermath of 9/11, the work has been interpreted as a response to the xenophobic atmosphere that ensued in New York. No single symbolic order dominates, but it may be interpreted that each symbolic order is in competition, alluding to the latent violence implicit in the politicization of religious belief. The work may also be seen to suggest that difference and connectivity are encoded at a cellular level.

Selected solo exhibitions

'Adel Abdessemed: Drawing for Human Park', Le Magasin-Centre National d'Art Contemporain, Grenoble, 2008
'Adel Abdessemed: Situation and Practice', MIT List Visual Arts Center, Cambridge, MA, 2008
'Adel Abdessemed: Don't Trust Me', Walter and McBean Galleries, San Francisco Art Institute, 2008
'Adel Abdessemed: Dead or Alive', P.S.1 Contemporary Art Center/ MoMA, Long Island City, NY, 2007
'Adel Abdessemed: Happiness in Mitte', Ikon Gallery, Birmingham, 2005

Selected group exhibitions

10th Havana Biennial, 2009
7th Gwangju Biennale, South Korea, 2008
'Flow', The Studio Museum in Harlem, New York, 2008
'Traces du Sacré', Centre Pompidou, Paris, 2008
'Think with the Senses, Feel with the Mind: Art in the Present Tense', 52nd Venice Biennale, Venice, 2007
'Notre Histoire', Palais de Tokyo, Paris, 2006

Selected publications

Farver, Jane et al. *Adel Abdessemed: Situation and Practice*, MIT Press, Cambridge, MA, 2008
Archer, Michael and Jenny Schlenzka, *Political/Minimal*, ed. Klaus Biesenbach, Verlag für moderne Kunst Nürnberg, Nuremberg, 2008
Heiss, Alanna and Neville Wakefield, *Adel Abdessemed: Dead or Alive*, P.S.1 Contemporary Art Center/MoMA, Long Island City, NY, 2007
Frogier, Larys et al. *Adel Abdessemed: A l'attaque*, JRP | Ringier, Paris, 2007

FRANZ ACKERMANN

Born in Neumarkt, Germany, 1963. Lives in Berlin, Germany.

Over the last twenty years, Franz Ackermann has become known for two major bodies of work: the large, brightly coloured installations featuring mounted sculptural constructions he has created between 2006 and the present; and his watercolour 'mental maps' created in the 1990s (paintings completed by Ackermann in hotel rooms recording a period of nomadic travel). His work often addresses the glamour of tourism and the sense of connectedness it imparts, as well as its underbelly – the disposability of its products and its environmental impact.

The diamond relief sculpture of *Roter Platz*, 2007, references carbon, the basis of natural life. But the literal translation of the work's title, 'red square', bears multiple meanings – the famous public square in Moscow, Russia, and geometric abstraction of modernist painting. The effect of the wall painting is cosmological in its scope, with the relief appearing like a satellite moving against a surreal, radically rendered universe.

Selected solo and two-person exhibitions

'Udomsak Krisanamis and a German Artist', Gavin Brown's enterprise, New York, 2007
'Home, Home Again, 23 Ghosts', White Cube, London, 2006
'Franz Ackermann', Irish Museum of Modern Art, Dublin, 2005
'Eine Nacht in den Tropen', Kunsthalle Nürnberg, Nuremberg, 2003
'The Waterfall', Museum of Contemporary Art, Chicago, 2002
'Seasons in the Sun', Stedelijk Museum, Amsterdam, 2002

Selected group exhibitions

'Compass In Hand: Selections from the Judith Rothschild Foundation Contemporary Drawings Collection', Museum of Modern Art, New York, 2009
'Altermodern: Tate Triennial 2009', Tate Britain, London, 2009
'Imagination Becomes Reality: Expanded Painting; Works From the Goetz Collection', ZKM Center for Art and Media, Karlsruhe, 2007
'Anstoss Berlin: Kunst macht Welt', Haus am Waldsee, Berlin, 2006
'Ecstasy: In and About Altered States', Museum of Contemporary Art, Los Angeles, 2005
'Monument to Now: The Dakis Joannou Collection', Deste Foundation for Contemporary Art, Athens, 2004
'Global Navigation System', Palais de Tokyo, Paris, 2003
'Hybrids', Tate Liverpool, 2001

Selected publications

Decter, Joshua et al, *Dark Places*, The Santa Monica Museum of Art, 2006
Gingeras, Alison M. et al, *The Triumph of Painting*, Saatchi Gallery, London, 2005
Cameron, Dan et al, *Monument to Now: The Dakis Joannou Collection*, ed. Jeffrey Deitch, Deste Foundation for Contemporary Art, Athens, 2004
Hoptman, Laura, *Drawing Now: Eight Propositions*, The Museum of Modern Art, New York, 2002
Frankel, David and Madeleine Grynstzein, *Carnegie International 1999/2000*, Carnegie Museum of Art, Pittsburgh, 1999

ANGELA BULLOCH

Born in Rainy River, Canada, 1966. Lives in Berlin, Germany.

The light installations and projections of Angela Bulloch's 'Night Sky' series depict twinkling night skies as they might appear from different locations in the universe, rather than from an earthly vantage point. By placing a technologically mediated presentation of the cosmological realm within a gallery context, *Night Sky: Mars from Venus.12*, 2008, positions a limitless universe in a finite environment. The viewer is potentially located in a double space – both in the gallery and in a representation of infinite space, between the intricate celestial construction and the reflective surface or mirrored universe adjacent to it.

Angela Bulloch's 'Night Sky' series has been developed concurrent to a body of work concerned with the legacies of modernist design and popular culture in relation to pixels and coding. Bulloch's 'Pixel Boxes' – sculptural boxes that appear as scaled-up and three-dimensional versions of a pixel, the minute coded elements of a digital image – formally reference modernism, including architecture and minimalism.

Selected solo exhibitions

"The Space That Time Forgot', Lenbachhaus, Kunstbau, Munich, 2008
'Angela Bulloch', De Pont Museum voor Hedendaagse Kunst, Tilburg, The Netherlands, 2006
'Angela Bulloch', The Power Plant, Toronto, 2006
'Angela Bulloch', Modern Art Oxford, 2005
'To the Power of 4', Secession Wien, Vienna, 2005
'Angela Bulloch', Le Consortium, Dijon, 2005
'New Work 8: Angela Bulloch, World Reflections', Aspen Art Museum, 2003
'Superstructure', Migros Museum für Gegenwartskunst, Zürich, 1998

Selected group exhibitions

'Theanyspacewhatever', Solomon R. Guggenheim Museum, New York, 2008
'Tate Triennial', Tate Britain, London, 2006
'Satellite Of Love', Witte de With, Rotterdam, 2006
'Extreme Abstraction', Albright-Knox Art Gallery, Buffalo, 2005
'Preis der Freunde der Nationalgalerie', Hamburger Bahnhof–Museum für Gegenwartskunst, Berlin, 2005
'Remix', Tate Liverpool, 2002
'Performance Anxiety', Museum of Contemporary Art, Chicago, 1997
'The Turner Prize 1997', Tate Gallery, London, 1997

Selected publications

Kittelmann, Udo, ed., *Angela Bulloch: Mud Bath*, The Green Box, Berlin, 2008
Gurk, Christoph and Axel Lapp, *Angela Bulloch: Prime Numbers*, Verlag der Buchhandlung Walther König, Cologne, 2006
Boase, Gair et al, *Tate Triennial 2006: New British Art*, eds. Beatrix Ruf and Clarrie Wallis, Tate Publishing, London, 2006
Eichler, Dominic, *Angela Bulloch*, Secession, Vienna, 2005
Grachos, Louis and Claire Schneider, *Extreme Abstraction*, Albright-Knox Art Gallery, Buffalo, 2005
Bourriaud, Nicolas, *Playlist*, Palais de Tokyo, Paris, 2004

MIRCEA CANTOR

Born in Oradea, Romania, 1977. Lives in Paris, France.

The video and mixed media installations of Mircea Cantor address issues of globalization and displacement by creating points of exchange where two or more worlds and ideas collide, producing subtle commentaries on contemporary life. Cantor's interests are suggested in the title of his recent exhibition 'The Need for Uncertainty'. They are also revealed in the work *Untitled*, 2006, which depicts a torn word mark of *Le Monde* with the addition of the letter 's' in red to the black and white logo to make 'Les Mondes', thereby multiplying uncertainty and suggesting plurality in conceptions of the universe. Similarly, the project commissioned for The Power Plant's *Switch* magazine depicts a performed activity where twelve young women, each born under a different zodiac sign, create the columns of the DNA double helix out of imprints of their lipstick. The columns repeat around four walls to form an enclosure – two opposing concepts of life forces linked through entrapment.

Airplanes and Angels, 2008, presents a flying carpet in suspended animation, a contradiction in itself. Although hand-woven by Romanian craftswomen using traditional techniques, the design depicting flying planes and soaring angels is not traditional. The work alludes to differing beliefs in the supernatural world, from angels to flying carpets, and meshes them against the plane as a symbol of nomadic life in a globalized world. On a more sinister note, the repetition of the planes on the carpet could be read as a military formation. The carpet is also inverted, adding to a sense of dislocation and uncertainty.

Selected solo exhibitions

'White Sugar for Black Days', Yvon Lambert, Paris, 2009
'The Need for Uncertainty', Camden Arts Centre, London, 2009
'Future Gifts', Mücsarnok Kunsthalle, Budapest, 2008
'The Title Is the Last Thing', Philadelphia Museum of Art, 2006
'Prix Ricard S.A.', Centre Pompidou, Paris, 2005

Selected group exhibitions

'Transmission Interrupted', Modern Art Oxford, 2009
'Here is Every: Four Decades of Contemporary Art', Museum of Modern Art, New York, 2008
'Brave New Worlds', Walker Art Center, Minneapolis, 2007
'Notre Histoire', Palais de Tokyo, Paris, 2006
'Universal Experience: Art, Life, and the Tourist's Eye', Museum of Contemporary Art, Chicago, 2005

Selected publications

Cotter, Suzanne, *The Need for Uncertainty* (cat.), Modern Art Oxford, 2009
Guillaume, Valerie and Christine Macel, *Airs de Paris* (cat.), Centre Pompidou, Paris, 2007
Buckley, Bernadette et al, *Paranoia* (cat.), Leeds Art Gallery, 2006
Isé, Claudine, *Displaced* (cat.), The Hammer Museum, Los Angeles, 2003

JANET CARDIFF & GEORGE BURES MILLER

Born in Brussels, Canada, 1957, and Vegreville, Canada, 1960.
Live in Berlin, Germany, and Grindrod, Canada.

Janet Cardiff and George Bures Miller are renowned for their multimedia, multi-sensory installations that utilize narrative structures and sound to alter perceptions of time and space. The piece *Telephone / Time*, 2004, presents a 1960s era office desk, chair and light with an outmoded analog telephone. Visitors who pick up the telephone overhear, but cannot connect to, a conversation between the artist, Janet Cardiff, and a male mathematician on the nature of time. Their conversation links theoretical cosmological phenomena, such as the event horizon, to the concept of time as linear and multi-dimensional, and time as felt experience, transformative and transitional, open to alteration by emotion, circumstance, logic, and illogic.

Selected solo exhibitions

'Opera for a Small Room', The Carnegie Museum of Art, Pittsburgh, 2009
'The Murder of Crows', Hamburger Bahnhof–Museum für Gegenwartskunst, Berlin, 2009
'Janet Cardiff: Forty-Part Motet', Moderna Museet, Stockholm, 2006
'Directions: Words Drawn in Water', Hirshhorn Museum and Sculpture Garden, Smithsonian Institution, Washington, D.C., 2005
'The Paradise Institute', National Gallery of Canada, Ottawa, 2002
'Janet Cardiff: Forty-Part Motet', P.S.1 Center for Contemporary Art/MoMA, Long Island City, NY, 2001

Selected group exhibitions

'Act/React: Interactive Installation Art', Milwaukee Art Museum, 2008–09
'Take Two: Worlds and Views: Contemporary Art from the Collection', The Museum of Modern Art, New York, 2005
'Ecstasy: In and About Altered States', The Museum of Contemporary Art, Los Angeles, 2005
'Fantasy Underfoot, The 47th Annual Corcoran Biennial', Corcoran Gallery of Art, Washington, D.C., 2003
'The Paradise Institute', The Canadian Pavilion, 49th Venice Biennale, 2001
'Carnegie International 1999/2000', The Carnegie Museum of Art, Pittsburgh, 1999–2000
'Present Tense: Nine Artists in the Nineties', San Francisco Museum of Modern Art, 1997

Selected publications

Fiona Bradley, ed., *Janet Cardiff & George Bures Miller: The House Of Books Has No Windows*, The Fruitmarket Gallery with Modern Art Oxford, 2008
Ralf Beil and Bartomeu Mari, eds., *Janet Cardiff & George Bures Miller: The Killing Machine and Other Stories 1995–2007*, Hatje Cantz Publishers with Museu d'Art Contemporani de Barcelona and Institute Mathildenhöhe Darmstadt, Ostfildern, Germany, 2007
Eccles, Tom and Francesca von Habsburg, *Janet Cardiff: The Walk Book*, TBA21, Vienna and Public Art Fund, New York, 2005
Christov-Bakargiev, Carolyn, *Janet Cardiff: A Survey of Works Including Collaborations with George Bures Miller*, P.S.1 Contemporary Art Center/MoMA, Long Island City, NY, 2002

CERITH WYN EVANS

Born in Llanelli, Wales, 1958. Lives in London, UK.

Cerith Wyn Evans is interested in intertextuality and often jumps across time to reconsider how developments in art, technology, science, and philosophy have influenced current understandings of the universe. The passage quoted in *With the advent of Radio Astronomy . . .*, 2004, is taken from the scientific text "Astro-photography Stages of Photographic Development," 1987 by Siegfried Marx. It refers to how the development of radio astronomy in the 1960s led to corrections in the previous photographic mapping of space. The minute dandruff flakes and emulsion flaws that led to the erroneous naming of galaxies can be likened to the immense challenge of mapping infinite space. Evans's implicit contrasting of infinite space with the microcosmic, in the form of specks of dust, can also be related to the Buddhist concept first recorded in the Buddhist text *Avatamsaka Sutra* – that the entire universe is present in a dust particle.[1] The Marx publication has been used by Evans as a source for a number of works, including *Cleave*, 2005, in which a World War II searchlight beams a section of text from the publication in Morse code vertically into the sky.

Transmission (John Cage), 2003, features a paper lantern, designed by Isamu Noguchi in the 1950s, which blinks a message in Morse code, a language developed in the nineteenth century with the advent of electricity and sometimes described as the first universal code. Evans uses this language, which became outmoded at the close of the twentieth century as the digital era took hold, to deliver John Cage's 'Lecture on Nothing (Silence)' first delivered by the avant-garde American composer around 1950. In keeping with the rhythm of the lecture, the computer-programmed lamp is inactive for periods, emulating the silences in the lecture. In the lecture, Cage says, "I have nothing to say . . . and I am saying it," a statement both negative and positive – a reflection perhaps of Cage's interest in indeterminacy and Zen Buddhism, and the idea that nothingness is a void of infinite possibilities.

Harmonia Mundi (firework), 2009, another text-based piece by Evans, is drawn from a quotation by Johannes Kepler from his book *Harmonices Mundi*, published in 1618. Kepler was a seventeenth-century German mathematician and astronomer whose theories and discoveries influenced astronomers and physicists such as Galileo and Newton. In Harmonices Mundi, Kepler relates the nature of the universe to harmonics. The piece was commissioned by The Power Plant and was ignited on the upper southwest exterior wall of the gallery on 11 June 2009 at 10 PM. Its lighting officially opened the 'Universal Code' exhibition.

The excerpt used by Evans for *Harmonia Mundi (firework)* read:
On the origin of the harmonic proportions,
And on the
Nature and differences
Of those things which are concerned with melody

NOTES

1. *Paul Williams,* Mahayana Buddhism: The Doctrinal Foundations. *(London: Routledge, 1989), 120–124.*

Cerith Wyn Evans contd.

Selected solo exhibitions
'Cerith Wyn Evans', Inverleith House, Edinburgh, 2009
'Bubble Peddler', Kunsthaus Graz, Austria, 2007
'Signatures of the Invisible', White Cube, London, 2007
'Cerith Wyn Evans', Museum of Fine Arts, Boston, 2004
'Cerith Wyn Evans', Centre Pompidou, Paris, 2004
'Cerith Wyn Evans: Cleave 00', Art Now series, Tate Modern, London, 2000

Selected group exhibitions
'Audio, Video, Disco', Kunsthalle Zurich, 2009
'Voice and Void', The Aldrich Contemporary Art Museum, Ridgefield, CT, 2007
'Tate Triennial 2006: New British Art', Tate Britain, London, 2006
'The Vanity of Allegory', Deutsche Guggenheim, Berlin, 2006
'Utopia Station', 50th International Venice Biennale, 2005
'Speaking of Sofas', Tate Gallery, London; Gavin Brown's enterprise, New York, 2003
'The Greenhouse Effect', Serpentine Gallery, London, 2000
'From the Corner of the Eye', Stedelijk Museum, Amsterdam, 1998
45th International Venice Biennale, 1995
'Image and Object in Current British Art', Centre Pompidou, Paris, 1990

Selected publications
Birnbaum, Daniel and Octavio Zaya, *Visibleinvisible*, Hatje Cantz, Munich, 2008
Budak, Adam et al, *Bubble Peddler*, ed. Peter Pakesch, Walther König and Kunsthaus Graz, Cologne, 2007
Boase, Gair et al, *Tate Triennial 2006: New British Art*, eds. Beatrix Ruf and Clarrie Wallis, Tate Publishing, London, 2006
Nesbitt, Molly et al, *Cerith Wyn Evans: In Which Something Happens All Over Again for the Very First Time*, Walther König/Lenbachhaus Munchen, 2006
Arning, Bill. *Thoughts Unsaid . . . Now Forgotten*, MIT List Visual Art Center, Cambridge, MA, 2004
Higgie, Jennifer, *Cerith Wyn Evans*, Camden Arts Centre, London, 2004
Meredith, Rachel, *Cerith Wyn Evans: Art Now*, Tate Publishing, London, 2000

HENRIK HÅKANSSON

Born in Helsingborg, Sweden, 1968. Lives in Varberg, Sweden, and Berlin, Germany.

For *Monarch–The Eternal*, 2008, Henrik Håkansson captures on film, in an unending loop, hundreds of Monarch butterflies after their mysterious annual migration from Ontario to Mexico. The film is the outcome of recordings shot with a static film camera on the Monarch butterfly reserves in Cerro Prieto, Michoacán, Mexico, on 22 February 2008. First presented in Mexico in 2008, the staging of *Monarch–The Eternal* in Toronto echoes the cycle of eternal return that sees the Monarch butterfly circle back to its northern summer feeding ground after four generations. The life cycle of the mature butterfly is about four to six weeks, except for the fourth generation – the one that flies to Mexico and lives up to eight months.

Henrik Håkansson records and reconstructs the life cycles of plants, birds and insects using media such as surveillance cameras, high-speed films and specialized computer programs, apparatuses more common to scientific experimentation than artistic research. By placing the natural world in an exhibition context, Håkansson fuses biological impulses with strategies of abstraction and objectification, creating the possibility for new dialogues between human beings and the natural world.

Selected solo exhibitions
'Cyanopsitta spixii Case Study #001', Isabella Stewart Gardner Museum, Boston, 2006
'Through the Woods to Find the Forest', Palais de Tokyo, Paris, 2006
'Henrik Håkansson', The Modern Institute, Glasgow, 2006
'An Introduction to the Birds', De Appel Centre for Contemporary Art, Amsterdam, 2003
'Tomorrow and Tonight', Kunsthalle Basel, 1999

Selected group exhibitions
'Radical Nature', Barbican Art Gallery, London, 2009
'Life Forms', Bonniers Konsthall, Stockholm, 2009
'Been Up So Long It Looks Like Down To Me', Presentation House Gallery, Vancouver, 2007
'Berlin North', Hamburger Bahnhof-Museum für Gegenwartskunst, Berlin, 2004

Selected publications
Porritt, Jonathan et al, *Radical Nature* (cat.), Barbican Art Gallery, London, 2009
Life Forms (cat.), Bonniers Konsthall, Stockholm, 2009
Miki, Akiko and Marc Sanchez, *Through the Woods to Find the Forest* (cat.), Palais de Tokyo, Paris, 2006
Heiser, Jörg and Annelie Pohlen, *An Introduction to the Birds* (cat.), De Appel Arts Centre, Amsterdam, 2003
Kempf, Helen Hirsch, *Henrik Håkansson* (cat.), Kunsthalle Basel, 1999

ANTONIA HIRSCH

Born in Frankfurt am Main, Germany, 1968. Lives in Vancouver, Canada.

Embedded within the gallery wall, the rhythmic sound of the installation *String Theory*, 2003, lures the viewer in to witness small flickers of movement. In a tiny, floor-level projection a figure skips haltingly, endlessly with a luminescent jump rope. The turning rope at times resembles the infinity symbol, or Möbius strip – continuously appearing and disappearing with each turn of the rope, flashes that can only be seen by viewers who crouch or bend to engage with the work.

The title, *String Theory*, alludes to a branch of theoretical physics concerning the nature of space-time. *String Theory* posits connections between differing mathematical and physical theorizations. At the heart of *String Theory* is the idea that the fundamental particles we observe are not zero-dimensional dots, but rather tiny one-dimensional strings so small that the most sophisticated instruments cannot indicate that they are not points. *String Theory* also predicts that there are extra dimensions to space beyond obvious length, breadth and depth, but we simply do not experience them. *String Theory* describes 10 to the power of 500 separate universes, all with different constants of nature and different laws of physics.

Selected solo exhibitions

'Anthropometrics Vol. I & Vol. II', Republic Gallery, Vancouver, 2009
'Photographie Metrique', Gallery 101, Ottawa, 2007
'Pulse', Artspeak, Vancouver, 2003
'Antonia Hirsch', La Centrale Galerie Powerhouse, Montreal, 2000

Selected group exhibitions

'How Soon is Now', Vancouver Art Gallery, 2009
'Friends of the Divided Mind', Royal College of Art Galleries, London, 2009
'Everything Should Be Made As Simple As Possible, But Not Simpler', Western Front, Vancouver, 2008
'75 Years of Collecting: Pop Icons', Vancouver Art Gallery, 2006
'The Knowledge', Norwich Art Gallery, UK, 1998

Selected publications

Rimmer, Cate, *The Banal*, SBC Gallery of Contemporary Art, Montreal, 2007
Cheng, Amy Huei-Hua, *Boundaries/Altered States*, Taipei Fine Arts Museum, 2006
Cheng, Amy Huei-Hua. *Ruins and Civilization*, Eslite Art Space, Taipei, 2004
Culley, Peter, *Antonia Hirsch. String Theory. A Note*, Xeno Gallery, Vancouver, 2003
Gale, Peggy, *Empire Line,* Gallery 44, Toronto, 2001; Toronto International Video Art Biennial, 2001

THOMAS HIRSCHHORN

Born in Bern, Switzerland, 1957. Lives in Paris, France.

Thomas Hirschhorn is internationally renowned for his installations, which feel like provisional spaces for social gatherings, such as pavilions or libraries. The makeshift appearance of his structures is shaped by cardboard, tape, tinfoil and other everyday materials, as well as pop culture imagery. For example, in his 'Universal Gym', 2009, exhibition at Gladstone Gallery, Hirschhorn constructed a non-functioning workout space, fastening barbells to the floor and replacing the useful armatures and components of weightlifting machines with cardboard tubes, tennis balls and plywood. While his installations have been interpreted as sharply critical of political situations, Hirschhorn often alludes to art, poetry and philosophy as being primary tools for understanding existence and bringing about change.

In the video installation *Dancing Philosophy*, 2007, Hirschhorn dances on four monitors, behind four different mannequins, with each monitor and mannequin depicting three twentieth-century philosophers – Antonio Gramsci, Georges Bataille and Gilles Deleuze – and one of philosophy's forefathers from the seventeenth century – Baruch Spinoza. When asked about *Dancing Philosophy* in an interview with *C Magazine*, Thomas Hirschhorn stated: "It is not important how you dance to philosophy. The important thing is to want to dance to it."[1]

NOTES

1. Fernando Oliva and Marcelo Rezende, "'I Believe the Solution is Panic': An Interview with Thomas Hirschhorn," C Magazine*; Spring 2007, 93; CBCA reference, 20*

Selected solo exhibitions

'Universal Gym', Gladstone Gallery, New York, 2009
'Jumbo Spoons and Big Cake', Musée d'Art Contemporain, Montreal, 2007
'Thomas Hirschhorn', Institute of Contemporary Art, Boston, 2005
'POLE-SELF', Centre Pompidou, Paris, 2001
'World Airport', The Renaissance Society, Chicago, 2000
'Focus', Art Institute of Chicago, 2000

Selected group exhibitions

'Carnegie International', Carnegie Museum of Art, Pittsburgh, 2008–09
'Unmonumental', New Museum, New York, 2008
'Airs de Paris', Centre Pompidou, Paris, 2007
'Universal Experience: Art, Life, and the Tourist's Eye', Museum of Contemporary Art, Chicago, 2005
'Dionysiac', Centre Pompidou, Paris, 2005
'Take Two: Worlds and Views, Contemporary Art from the Collection', The Museum of Modern Art, New York, 2005
'Common Wealth', Tate Modern, London, 2003
'Protest & Survive', Whitechapel Art Gallery, London, 2000
'Unfinished History', Walker Art Center, Minneapolis, 1998

Selected publications

Bonami, Francesco and Julie Rodrigues, *Universal Experience: Art, Life, and the Tourist's Eye*, ed. Tricia Van Eck, Museum of Contemporary Art, Chicago, 2005
Buchloh, Benjamin D.H. et al, *Thomas Hirschhorn*, Phaidon, London, 2004
Rondeau, James et al, *Jumbo Spoons and Big Cake*, The Art Institute of Chicago; World Airport, The Renaissance Society at The University of Chicago, 2000
Fleck, Robert, Alison Gingeras, Philippe Vergne, and Marcus Steinweg. *Doug Aitken, Nan Goldin, Thomas Hirschhorn*, Parkett Publishers, New York, 1999
Gingeras, Alison. *Thomas Hirschhorn*, Chisenhale Gallery, London, 1998

ANN VERONICA JANSSENS

Born in Folkstone, UK, 1965. Lives in Brussels, Belgium.

Ann Veronica Janssens is interested in natural phenomena, abstraction and intangibility. Her works frequently challenge the thresholds of vision and spatial perception, as devices designed to assist viewers to experience light, colour, sound, time, and space in different ways. In Janssens' words, quoted in an interview with Hans Theys, "nothing is more beautiful than a person's own perception. I try to push it to its limits."[1] Janssens frequently employs coloured light in her projection-based works and video installations, and as a means of dissolving forms to liberate architectural space and generate immersive, kinaesthetic viewing experiences. *Blue, Red and Yellow*, 2001, installed on the terrace of the Neue Nationale Galerie Berlin, featured an enclosure of translucent pavilion walls covered with transparent blue, red and yellow films, where the spectator could advance through misty, dematerialized, abstraction.

Side (Natural Version), 2006, is a video of a total eclipse of the sun, filmed with Guillaume Bleretin in Side, Turkey, on 29 March 2006. Solar eclipses typically feature light too intense for the human eye to modulate by revealing the outer layer of the sun's atmosphere. Eclipses have been known to cause temporary or permanent blindness, and Janssens' works refers indirectly to the optical scientific experiments of nineteenth-century Belgian physicist Joseph Plateau (1801–1883).[2] Plateau, inventor of the phenakistoscope, was interested in the psycho-optical principle of persistence of vision – the eye's momentary retention of an image after the object is no longer visible. Plateau is rumoured to have lost his sight to solar retinitis later in life as a result of performing experiments in which he stared into the sun for a full twenty-five seconds.[3]

Side (Studio Version), 2006, produced by Wiels, Centre for Contemporary Art, Brussels, differs from *Side (Natural Version)* through its transformative use of colour. *Side (Studio Version)* depicts an eclipse of the sun in a modulating series of colours, including green, blue, black, magenta, yellow, purple, red, and white.

Janssens' monitor piece *Crépuscule à Jantar Mantar (Vénus)*, 2008, documents the ancient building Jantar Mantar, a collection of architectural astronomical instruments built between 1727–34 by (then ruler) Maharaja Jai Singh II. The observatory in Jaipur, India, featured in Janssens' film, holds fourteen major geometric devices for measuring time, predicting eclipses, ascertaining star and planetary orbits, and determining the celestial altitudes. Local students of astronomy and Vedic astrology in Jaipur are required to take some of their lessons at the observatory. The observatory is considered spiritually and culturally significant, as the single most representative surviving work of Vedic thought.[4] Janssens' camera work in *Jaiprakash Yantra*, 2008, never clearly depicts the true scale of Jantar Mantar's instruments as the observatory tools are shot in detail such that their purpose is obfuscated and uncertain. In Crépuscule à Jantar Mantar (Vénus), the tools for celestial observation at Jantar Mantar are recorded from dawn to dusk to capture varying qualities of light, evoking a sense of mysticism.

NOTES

1. *http://www.hanstheys.be/artists/ann_veronica_janssens*
2. *http://www.argosarts.org*
3. *J. Plateau, Bibliographie analytique des principaux phénomènes subjectifs de la vision (Mém. Acad. Roy. Belg. XLII, 1°(1–59), 2°(1–59), 3° (1–26.), 4°(1–44.), 5°(1–35), 6° (1–45), Suppl. 1877, 1880 et 1882); cited by August Meessen, in "Apparitions and Miracles of the Sun," 3.*
4. *http://www.jantarmantar.org*

Selected solo and two-person exhibitions

'Are You Experienced?', Espai d'Art Contemporani de Castelló, Valencia, 2009
'Experience of Color–Ann Veronica Janssens and Diana Thater', Amherst Fine Arts Center, 2008
'Experiences and Sketches', Esther Schipper, Berlin, 2008
'Ann Veronica Janssens', Micheline Szwajcer, Antwerp, 2008
'Avril', Air de Paris,, 2006
'Ann Veronica Janssens', IKON Gallery, Birmingham, 2002
'Light Games', Neue Nationalgalerie, Berlin, 2001

Selected group exhibitions

'Faux-Jumeaux', Stedelijk Museum voor Actuele Kunst, Ghent, Belgium, 2008
'Fragile Beauty', Kunst Palast, Düsseldorf, 2008
'Stay for ever and ever and ever', South London Gallery, 2007
'Space Boomerang', Swiss Institute, New York, 2006
'Ecstasy: In and About Altered States', The Museum of Contemporary Art, Los Angeles,, 2005
'Eyes, Lies and Illusions', Hayward Gallery, London, 2004
'The World on its Head', San Francisco Art Institute, 2000

Selected publications

Francois, Michel, *Experienced*, Espai d'Art Contemporani de Castello, Base Publishing, Valencia, 2009
Bal, Mieke et al, *An den Fruhling*, ed. Markus Heinzelmann, Museum Morsbroich, Leverkusen, Germany, 2007
Christov-Bakargiev, Carolyn et al, *Ecstasy: In and About Altered States*, ed. Lisa Mark, The Museum of Contemporary Art, Los Angeles, 2005
Bal, Mieke, *Ann Veronica Janssens: Lichtspiel, Kunstverein Munchen*, Berliner Kunstlerprogramm, DAAD, Berlin, 2001
Van Leeuw, Marianne, *Ann Veronica Janssens*, Museum van Hendendaagse Kunst Antwerpen, Belgium, 1997

KIMSOOJA

Born in Taegu, South Korea, 1957. Lives in New York, USA.

Mandala: Zone of Zero, 2003, mixes Western popular culture with Eastern religious tradition. The room installation is illuminated by three sculptural shapes on three walls. Although these shapes are reminiscent of Tibetan mandalas – symbols for the design of the universe – they are jukebox speakers brimming with bubbles. Each speaker on the gallery walls play a different form of chanting – Tibetan, Islamic and Gregorian – songs with histories which span thousands of years. The immersive formal qualities of the sound and sculptural elements of the installation continue an interest in enveloping present in many of Kimsooja's works, such as installations that utilize Ybulbo – the Korean word for the woven fabrics bundled and used for a variety of purposes such as sleep, travel and the safekeeping of objects.

Kimsooja pursues in her work a sense of universality that is timeless and related to human experience. *Mandala: Zone of Zero* explores ideas of oneness and totality, while also highlighting distinctness and similarity in different spiritual traditions.

Selected solo exhibitions

'Black Box: Kimsooja', Hirshhorn Museum and Sculpture Garden, Washington, 2008
'Lotus: Zone of Zero', Rotonde Ravenstein, Korean Art Festival BOZAR, Brussels, 2008
'To Breathe: Invisible Mirror/Invisible Needle', Théâtre Chatelet, Nuit Blanche, Paris, 2008
'Seven Wishes and Secrets', MIT List Visual Arts Center, Cambridge, MA, 2005
'A Needle Woman', P.S.1 Contemporary Art Center/MoMA, New York, 2001

Selected group exhibitions

'It's Your Bright Future: 12 Korean Artists', Los Angeles County Museum of Art, 2009
'Against Exclusion: 3rd Annual Moscow Biennale', The Garage, Moscow, 2009
'Self/Other', The National Museum of Modern Art, Tokyo, 2008
'Critical Mass', Kunsthalle Bern, 2007
'Always a Little Further', 51st Venice Biennale, 2005

Selected publications

Bourgeois, Caroline, *Un Certain État du Monde?: A Selection of Works from the Francois Pinault Collection*, Skira, Milan, 2009
Otani, Shogo, *Self/Other* (cat.), The National Museum of Modern Art, Tokyo, 2008
Bussard, Katherine A. et al, *Street Art, Street Life* (cat.), eds. Lydia Lee and Whitney Rugg, Bronx Museum of the Arts, New York, 2008
Chadwick, Whitney, 'Worlds Together, Worlds Apart', *Women, Art and Society*, 4th Edition, Thames & Hudson, London, 2007

JED LIND

Born in Toronto, Canada, 1978. Lives in Los Angeles, USA.

The photographic series 'Better to Burn Out' results from carefully timed exposures of the night sky. Shot over one year in Joshua Tree, Idyllwild and Mount Pinos, in and around the new moon (the darkest part of the moon cycle), Jed Lind performed different processes to create each work in the series. For some photographs, Lind shifted his camera in and out of focus over the course of half an hour. For others, he pointed the camera into the night sky and slowly rotated the camera mounted on a tripod for hours. *Better to Burn Out No.1*, 2008, was created using a medium format camera to follow the trail of a single star. The circles in *Better to Burn Out No.1* result from the rotation of the earth, while the actual lines seen were drawn through the individual rotations of the camera. Lind likens these lines to the grooves in vinyl records which transmit sound. He characterizes the photograph *Better to Burn Out No.1* as the "result of slow and timely revolutions, a walking mantra."[1]

References for *Better to Burn Out No.1* include the 1890s 'celestographs' produced by August Strindberg. Strindberg was said to be distrustful of the lens apparatus and eliminated the use of the camera to capture images, instead opting to expose his plate glass film directly to the sky. Strindberg's resultant images were abstractions resembling the night sky.

NOTES

1. Jed Lind, Letter to the Power Plant, 2 June 2009.

Selected solo exhibitions

'Fluid Geographies', Jessica Bradley Art and Projects, Toronto, 2008
'A False Dawn', Jessica Bradley Art and Projects, Toronto, 2005
'Domes and Guiding Lights', MFA Thesis Show, D301 Gallery, California Institute of the Arts, Valencia, 2004
'Mirrored Circle', Mint Gallery, California Institute of the Arts, Valencia, 2003

Selected group exhibitions

'Infrastructure', Wignall Museum, Los Angeles, 2008
'Incognito', Santa Monica Museum of Art, Los Angeles, 2007
'Notes from the Overpass–St Petersburg–Los Angeles', Gallery G18, Helsinki, 2007
'Supersonic: 1 wind tunnel, 8 schools, 128 artists', Art Center College of Design, Pasadena, 2004

Selected publications

Rogers, Brett Cody, *Supersonic* (cat.), Art Center College of Design, Pasadena, 2004

IÑIGO MANGLANO-OVALLE

Born in Madrid, Spain, 1961. Lives in Chicago, USA.

Iñigo Manglano-Ovalle explores social issues such as illegal immigration, xenophobia and climate change in his video installations and sculptures, using motifs such as icebergs, clouds and other terrestrial and extra-terrestrial phenomena. He frequently collaborates with geneticists, biotech researchers and medical ethicists to create his work as a means of exploring the role of science in culture. In *Wind Shear*, 1997, surveillance technology typically reserved for monitoring political borders is inverted. Cameras in Nogales, Arizona, and Nogales, Sonora, Mexico, were pointed upwards to record the limitless sky hovering between the borders of the United States and Mexico. The recording alludes to the meteorological event of wind shear, as clouds move in opposition to each other, north and south simultaneously, at different elevations. The clouds depicted in *Wind Shear* appear first as nomadic bodies, ephemeral and free, yet also appear to be turned back, due to the intermittent reversal of the film and thereby mirroring the process of border control.

Selected solo exhibitions

'Iñigo Manglano-Ovalle', Donald Young Gallery, Chicago, 2008
'Iñigo Manglano-Ovalle', Galerie Thomas Schulte, Berlin, 2007
'Focus: Iñigo Manglano-Ovalle', The Art Institute of Chicago, 2005
'White Flags', Barcelona Pavilion, Mies van der Rohe Foundation, Barcelona, 2002
'Iñigo Manglano-Ovalle', Cleveland Center for Contemporary Art, 2001
'Balsero', Museum of Contemporary Art, Chicago, 1997

Selected group exhibitions

'Human|Nature: Artists Respond to a Changing Planet', Museum of Contemporary Art, San Diego; Berkeley Museum of Art/Pacific Film Archive, 2008
'Dokumenta XII', Kassel, Germany, 2007
'Weather Report: Art and Climate Change', Boulder Museum of Contemporary Art, 2007
'What's New Pussycat?' Museum fur Modern Kunst, Frankfurt, 2005
'Moving Pictures', Solomon R. Guggenheim Museum, New York, 2002
'Tempo', The Museum of Modern Art, New York, 2002
'What's New: Recent Acquisitions in Photography', Whitney Museum of American Art, New York, 2001
'Interventions: New Art in Unconventional Spaces', Milwaukee Art Museum, 2000

Selected publications

Ohlenshlager, Karin, *Ecomedia: Ecological Strategies in Today's Art*, Hatje Cantz Verlag, Berlin, 2007
Barliant, Claire and Kris Douglas, *Iñigo Manglano-Ovalle: Blinking Out of Existence*, Rochester Art Center, 2007
Lippard, Lucy R. et al, *Weather Report: Art and Climate Change*, Boulder Museum of Contemporary Art in collaboration with EcoArts, 2007
Binstock, Jonathan P. and Stacey Schmidt, *Closer to Home: The 48th Corcoran Biennial*, The Corcoran Gallery of Art, Washington, D.C., 2005
Hanhardt, John G. and Nancy Spector, *Moving Pictures: Contemporary Photography and Video from the Guggenheim Museum Collections*, The Solomon R. Guggenheim Foundation, New York, 2003

JOSIAH MCELHENY

Born in Boston, USA, 1966. Lives in Brooklyn, USA.

Josiah McElheny's intricate glass-blown sculptures and installations are often based around historical events and seek to fuse materiality and thought in the experience of looking. McElheny is interested in refection as a physical event and a conceptual process, the craft of glassmaking, and legacies of conceptual art.

For the body of work *Island Universe*, 2008 – which consists of several glass-blown sculptures and a film installation – McElheny sought a collaboration with cosmologist David Weinberg. The sculptures of *Island Universe* are scientifically accurate models of stages in the Big Bang theory: the varying lengths of rods are based on time measurements; clusters of glass discs and spheres accurately represent the clustering of galaxies, and light bulbs mimic quasars, the universe's brightest objects. The starting point for the sculptures was the opulent chandeliers of the New York Metropolitan Opera House, designed by Lobmeyr in 1965, the same year that the first data supporting the Big Bang theory was made public.

The film *Island Universe* was made in association with the larger body of work. The film was recorded on location at the Metropolitan Opera House and features the Lobmeyr chandeliers. It is divided into five sections that depict five types of universes. Camera movements, edits and sounds are choreographed in relation to scientific speculations about the origins of the universe. The film's commissioned soundtrack was composed by Paul Schütze.

Selected solo exhibitions

'Josiah McElheny', Andrea Rosen Gallery, New York, 2009
'Island Universe', White Cube, London; Museo de Art Reina Sofia, Madrid, 2008–09
'Projects 84: Josiah McElheny', The Museum of Modern Art, New York, 2007
'The Alpine Cathedral and The City-Crown', Moderna Museet, Stockholm, 2007
'Josiah McElheny', The Isabella Stewart Gardner Museum, Boston, 1999

Selected group exhibitions

'Super Vision', Institute of Contemporary Art, Boston, 2006
'Shiny', Wexner Center for the Arts, Columbus, OH, 2006
'Extreme Abstraction', Albright-Knox Art Gallery, Buffalo, 2005
'Beau Monde: Toward a Redeemed Cosmopolitanism', SITE Santa Fe, 2001
'Whitney Biennial: 2000', The Whitney Museum of American Art, New York, 2000
'Young Americans: Part II', Saatchi Gallery, London, 1998

Selected publications

Burnett, Craig et al, *Island Universe*, Jay Jopling/White Cube, London, 2008
Weinberg, David et al, *Notes for a Sculpture and a Film*, ed. Helen Molesworth, Wexner Center for the Arts, The Ohio State University, Columbus, 2006
Apollonia Morrill, ed., *Josiah McElheny: The Metal Party*, The Public Art Fund, New York; Yerba Buena Center for the Arts, San Francisco, 2002.

TANIA MOURAUD

Born in Paris, France, 1942. Lives in Paris, France.

The visceral installation *La Fabrique,* 2006, depicts multiple workers absorbed in the repetitive activity of weaving, accompanied by a clattering, cacophonous and ultimately rhythmic soundtrack, as the sounds from each loom are layered upon each other. Shot on location in a carpet-weaving facility in Kerala, India, the installation enacts a tension between the individual and the idea of connectedness through collective humanity. In ancient mythology, weaving is a metaphor for the origins of the universe, while in many cultures weaving is a means to create forms that signify the interconnectedness of life. On entering the installation the viewer is surrounded by multiplicity and a sense of connectedness, yet as a spectator the viewer is drawn to and confronted by the individuality and alterity of the Indian workers.

A truism of the information age is that it is post-industrial. During the industrial era theorists often figured the body as a machine. Following the growth in the ubiquity of information technologies coupled with research into DNA coding and artificial intelligence, it is becoming increasingly common to figure the body as information. *La Fabrique* reframes our considerations of globalization and technological advance by reminding us that manual labour continues to be the mainstay of production in many societies.

Selected solo exhibitions

'Ad Infinitum', Musée des Beaux-Arts de Nantes, 2009
'Borderland', Galerie Dominique Fiat, Paris, 2009
'Roaming Borderland', Musée de la Chasse et de la Nature, Paris, 2008
'Wall Painting', The Hammer Museum, Los Angeles, 1999
'World Signs II', Riverside Studios, London, 1998

Selected group exhibitions

'elles@pompidou', Centre Pompidou, Paris, 2009
'Blind Chance & Possible Futures', Nieuwe Vide, Haarlem, The Netherlands, 2009
'!Revolution?', Müscarnok Kunsthalle, Budapest, 2007
'Sous le manteau', Galerie Thaddeus Ropac, Paris, 1997
'Crossroad', Centre d'Art Contemporain, Brussels, 1996

Selected publications

Egana, Miguel, *Periphérique Intérieur*, Editions L'Harmattan, Paris, 2008
Kremeier, Ulrike et al, *!Revolution?* (cat.), ed. Livia Paldi, Müscarnok Kunsthalle, Budapest, 2007
Gonnard, Catherine and Elisabeth Lebovici, *Femmes Artistes/Artistes Femmes*, Editions Hazan, Paris, 2007
Morgan, Robert C., 'Tania Mouraud and the Paradox of Language', *Between Modernism and Conceptual Art: A Critical Response*, McFarland & Co., Jefferson, NC, 1997

GABRIEL OROZCO

Born in Jalapa, Mexico, 1962. Lives in New York, USA, Paris, France, and Mexico City, Mexico.

Gabriel Orozco's art practice emerges from his nomadic lifestyle. Working without a studio, his sculptures and photographs range from subtle interventions to spectacular alterations of found objects. Orozco has an enduring interest in games as a model for cultural processes. In an interview with the PBS television show *Art:21*, Orozco states: "I think every game is a universe in a way, or every game is an expression of how the universe works, for different cultures . . . I like to make a game grow by trying to understand the geometry of the situation"[1]

For the sculpture *Black Kites*, 1997, Orozco carefully drew over the surface of a human skull using graphite pencil, until the full surface yielded a checkerboard, or harlequin, pattern. The skull, as a symbol, is entrenched in the Western art historical tradition of the 'memento mori' or 'vanitas' as well as the Mexican Day of the Dead carnival. The translation of *Black Kites* from one medium to another — from sculpture to photography, with *Black Kites Perspective (front horizontal)*, 1997 — could be seen to further emphasize the polarities suggested by the piece: the organic versus the technical, and the natural versus the manufactured.

NOTES

1. *http://www.pbs.org/art21/artists/orozco/clip1.html*

Selected solo exhibitions

'Gabriel Orozco', The Museum of Modern Art, New York; Kunstmuseum Basel; Centre Pompidou, Paris; Tate Modern, London, 2009–11
'Gabriel Orozco', White Cube Gallery, London, 2006
'Gabriel Orozco: Photographs', Hirshhorn Museum, Washington, D.C.; Serpentine Gallery, London, 2004
'Gabriel Orozco', The Museum of Contemporary Art, Los Angeles, 2000
'Recordings and Drawings', Stedelijk Museum, Amsterdam, 1997
'Gabriel Orozco', Museum of Contemporary Art, Chicago, 1994

Selected group exhibitions

'Hidden in Plain Sight: Contemporary Photographs from the Collection', The Metropolitan Museum of Art, New York, 2007
'Part Object, Part Sculpture', Wexner Center for the Arts, Columbus, OH, 2006
'Tempo', The Museum of Modern Art, New York, 2002
'Sonic Process', Centre Pompidou, Paris, 2002
'Carnegie International', Carnegie Museum of Art, Pittsburgh, 1999
'In Transit', The New Museum of Contemporary Art, New York, 1993

Selected publications

Jay Jopling, ed., *Gabriel Orozco*, White Cube, London, 2006
Fer, Briony et al, *Gabriel Orozco*, Serpentine Gallery, London, 2004
Fineman, Mia and Phyllis Rosenzweig, *Gabriel Orozco: Photographs* (cat.), Hirshhorn Museum and Sculpture Garden, Smithsonian Institution, Washington, D.C.; Steidl, Göttingen, Germany, 2004
Buchloh, Benjamin et al., *Gabriel Orozco* (cat.), The Museum of Contemporary Art, Los Angeles, 2000

THE OTOLITH GROUP

Founded in London, UK, by Anjalika Sagar, Kodwo Eshun and Richard Couzins, 2002. Based in London, UK.

Otolith, 2003, is a hybrid of science fiction and documentary filmmaking. The film is set in the twenty-second century, when a community of humans are effectively extraterrestrials, having lost the ability to withstand the earth's gravitational force. The film utilizes archival footage to reflect on possible alternative futures of the recent past. Dr. Usha Adelbaran Sagar (a fictional descendant of artist member Anjalika Sagar of The Otolith Group) narrates the experiences of several generations of women of the South Asian Sagar family. She links her history to Anjalika Sagar's grandmother, who was president of the National Federation of Indian Women and who, during the 1960s, met Russian cosmonaut Valentina Tereshkova, the first woman to orbit the earth. The film alludes to the false optimism of the pioneering era of space travel and includes footage of Sagar, the artist, floating in microgravity at Star City, the former Soviet Aerospace facility, where cosmonauts are still trained.

Otolith explores the potential value of archival footage and evokes the poetry of mediated memory, utilizing 35mm images of the nuclear mushroom clouds that have permanently marked collective memory; digital video footage of the London peace marches prior to the Iraq war; footage from the World War II era of non-alignment, and private memories shot on 8mm film in 1970s London and Haridwar.

Selected solo exhibitions

'A Long Time Between Suns', Gasworks, London, 2009
'The Otolith Group', Argos, Brussels, 2007

Selected group exhibitions

'Translocalmotion: 7th Shanghai Biennale', 2008
'Italia italie italien italy wlochy', ARCOS–Museo di Arte Contemporanea del Sannio, Benevento, 2008
'1st Athens Biennial', 2007
'You have not been honest', Museo D'Arte Contemporanea Donna Regina, Naples, 2007
'2° Bienal Internacional de Arte Contemporáneo de Sevilla–BIACS–Fundación Bienal Internacional de Arte Contemporáneo de Sevilla', 2006
'Tate Triennial–New British Art', Tate Triennial, London, 2006

Selected publications

Demose, TJ and Irmgard Emmelhainz, *A Long Time Between Suns*, Sternberg Press and Gasworks, London, 2009

ADRIAN PACI

Born in Shkoder, Albania, 1969. Lives in Milan, Italy.

The single photograph *Per Speculum*, 2006, by Adrian Paci relates to his film of the same name. The film begins by depicting a seemingly timeless and mythological scene of children caught in the refection of a mirror and ends with the children in a tree holding shards of the mirror, reflecting the sun. The allegorical film contains references to fairy tales, the Bible, and art and film history. The phrase 'Per Speculum' translates as "by means of a mirror."

In the photograph we see a tree, an ancient symbol for life. A child sits reflecting the sun, another symbol of life, surrounded by a corona that metaphorically links the child to the universal. The photograph references the Bible's First Letter to the Corinthians, which states "At present we see indistinctly, as through a mirror, but then face to face. At present I know partially; then I shall know fully, as I am fully known."[1] In his paintings, films, photographs, performances, and sculptures, Paci typically uses narrative strategies to consider his personal experiences of exile. In *Per Speculum* he draws upon allegory to ponder connectedness and fragmentation through the disruption of a child's utopian world.

NOTES

1. *Paul's First Letter to the Corinthians (13:12)*

Selected solo exhibitions

'Per Speculum', Milton Keynes Gallery, Milton Keynes, UK, 2007
'Adrian Paci', P.S.1 Contemporary Art Center/MoMA, Long Island City, NY, 2005
'Perspective 147: Adrian Paci', Contemporary Arts Museum Houston, 2005
'Adrian Paci', Moderna Museet, Stockholm, 2005
'A Toll on Rituals', Baltic Art Center, Gotland, Sweden, 2003

Selected group exhibitions

'Panoramica', Museum Tamayo, Mexico City, 2009
'Street & Studio: An Urban History of Photography', Tate Modern, London, 2008
'Shifting Identities', Kunsthaus Zurich, 2008
'Auto Emotion: Autobiography, emotion and self-fashioning', The Power Plant, Toronto, 2007
'Senso Unico', P.S.1 Contemporary Art Center/MoMA, Long Island City, NY, 2005
'Busan Biennale', Korea, 2006
'New Video, New Europe', The Renaissance Society, Chicago, 2004

Selected publications

Mashenson, Edna, *Adrian Paci: Subjects in Transit* (cat.), Center for Contemporary Arts, Tel Aviv, 2008
Engler, Martin et al, *Adrian Paci*, Kunsteverein Hannover, 2008
Stanely, Michael et al, *Adrian Paci Per Speculum* (cat.), Milton Keynes Gallery, Milton Keynes, UK, 2007
De Corral, Maria and Rosa Martinez, *Always a Little Further*, 51st Venice Biennale, Rizzoli International Publications, New York, 2005
Bonami, Francesco, Exit, *nuove geografie della creatività italiana*, Milan, 2003

TREVOR PAGLEN

Born in Maryland, USA, 1974. Lives in Oakland, USA.

Trevor Paglen is an artist, writer and experimental geographer. In past projects, Paglen has explored and documented hidden military landscapes. For *Code Names*, 2007, Paglen investigated and listed the thousands of phrases that designate active military programs with classified purposes. Code names recorded include 'Ontario Water', 'Science Dust', 'Project 999', 'Service Star', 'Seven Citadels', and absurdly, 'Sly Stone'. For his ongoing project 'The Other Night Sky', Paglen is tracking and photographing every classified American satellite in Earth's orbit – nearly two hundred covert spacecraft.

Active Military and Reconnaissance Satellites of the United States of America, 2008, shows the paths of these tracked satellites, projected onto a scale model of the planet Earth, accompanied by a soundtrack of intercepted satellite transmissions. The data plotting the paths of the satellites was supplied by an international network of amateur satellite observers, while the silk-screened sculpture of the earth tracked by military satellites evokes a war room scenario. In developing the work, Paglen was inspired by the methods of early astronomers like Johannes Kepler and Galileo Galilei, who documented previously unseen moons of Jupiter in the early seventeenth century. Paglen figures these satellites as the unseen or unacknowledged moons of the earth.

Selected solo exhibitions

'Trevor Paglen', Bellwether Gallery, New York, 2009
'The Other Night Sky: Matrix 225', Berkeley Art Museum, 2008
'Black World', Bellwether Gallery, New York, 2006
'Trevor Paglen', The LAB, San Francisco, 2005
'Trevor Paglen', Deadtech, Chicago, 2001

Selected group exhibitions

'Experimental Geography', Richard E. Peeler Art Center, De Pauw University, Greencastle, IN, 2008; Colby College Museum of Art, Waterville, ME (upcoming)
'The SECA Art Award Exhibition', San Francisco Museum of Modern Art, 2009
'Berkeley Big Bang 08', UC Berkeley Art Museum and Pacific Film Archive, 2008
'6 Billion Perps Held Hostage! Artists Address Global Warming', The Andy Warhol Museum, Pittsburgh, 2007

Selected publications

Trevor Paglen, Aperture Foundation, New York, 2009
Thompson, Nato & Independent Curators International, *Experimental Geography*, Melville House Publishing, Brooklyn, 2009
Paglen, Trevor, *Blank Spots on the Map: The Dark Geography of the Pentagon's Secret World*, Penguin Publishers, New York, 2009
Keenan, Thomas, 'Disappearances: The Photographs of Trevor Paglen', *Aperture* 191, Summer 2008
Paglen, Trevor. *I Could Tell You but Then You Would Have to Be Destroyed by Me: Emblems from the Pentagon's Black World*, Melville House Publishing, Brooklyn, 2007

KATIE PATERSON

Born in Glasgow, Scotland, 1981. Lives in London, UK.

In recent works Katie Paterson has been interested in the disappearance and loss of earthly and cosmological phenomena, such as dead stars and melting glaciers. For *Earth-Moon-Earth (Moonlight Sonata Reflected from the Surface of the Moon)*, 2007, Paterson translated Beethoven's 'Moonlight Sonata' into Morse code and sent it to the moon via E.M.E. – Earth-Moon-Earth, also known as moon bounce – a radio communications technique. Bounced off the moon, the coded score returned to earth fragmented by the moon's surface. Paterson retranslated the composition into a new score, including the missing notes. The score is included in the installation, as is the coded transmission, which can be heard through headphones. The centrepiece though is the computer-programmed player piano. It renders the familiar strains of the Beethoven composition, but the melody contains errors, missing notes and incomplete phrases – losses incurred due to the uneven surface of the moon and imperfect reflection of the code over its 384,000-kilometre journey. The work is haunted by absence, distance and loneliness.

Selected solo and two-person exhibitions

'Earth-Moon-Earth' with David Lamelas, Lakeside Centre, Nottingham, 2009
'Encounters: Katie Paterson', Modern Art Oxford, 2008
'Katie Paterson', Albion Gallery, London, 2008
'Langjökull, Snæfellsjökull, Solheimajökull', ROOM Artspace, London, 2008
'Katie Paterson', Matthew Brown Gallery, London, 2007

Selected group exhibitions

'Life Forms', Bonniers Konsthall, Stockholm, 2009–10
'Performa 09', Performa, New York, 2009
'Mutisme(s)', Lieu-Commun Espace D'Art Contemporain, Toulouse, 2009
'Contemporary Eye: Material Matters', Pallant House, West Sussex, 2009
'Altermodern: Tate Triennial', Tate Britain, London, 2009

Selected publications

Nicolas Bourriaud, ed., *Altermodern: Tate Triennial 2009*, Tate Publishing, London, 2009
Paterson, Katie, *Earth-Moon-Earth (Moonlight Sonata Reflected from the Surface of the Moon)*, Modern Art Oxford, 2008

FRED TOMASELLI

Born in Santa Monica, USA, 1956. Lives in Brooklyn, USA.

Fred Tomaselli works in two-dimensional mediums such as resin, photogram, watercolour, tapestry, and newsprint collage. Tomaselli's large, colour-saturated collages are hallucinogenic, kaleidoscopic and surreal, frequently drawing on imagery and illustrations from scientific, decorative and subcultural traditions, including diagrams from theoretical physics, Tibetan mandalas, and psychedelic and folk art. His references are often simultaneously microcosmic and macrocosmic and built of collaged layers, such as his paintings of narcotic star maps of constellations.

Car Bomb, 2008, depicts a cosmic vortex radiating from a pyre of exploding cars. It is a scene of atomization and hybridization, with elements connected as decorative strings of automobile and body parts, ornamental motifs and eyes. The piece has multiple references, from the stylized sun sets of art deco, to DNA and binary code, to cosmological theories of the Big Bang and the violence associated with jihadist fervour. In the catalogue *Fred Tomaselli: Monsters of Paradise*, Tomaselli says he is concerned with infection, pathology, pain, and pollution. He notes that we "see the world through a scrim of ideologies and technologies and the crackling static of chemicals and electronic media."[1]

NOTES

1. Fred Tomaselli: Monsters of Paradise, *Edinburgh, 2004, 43.*

Selected solo exhibitions

'Fred Tomaseli', Aspen Art Museum, CO; Tang Museum, Skidmore College, Saratoga, NY; Brooklyn Museum of Art, 2009
'Fred Tomaselli', Albright-Knox Gallery of Art, Buffalo, 2003
'Gravity's Rainbow', Whitney Museum of American Art at Philip Morris, New York, 1999
'Fred Tomaselli', White Columns, New York, 1991

Selected group exhibitions

'Prospect.1 New Orleans', 2008–09
'Ecstasy: In and About Altered States', Museum of Contemporary Art, Los Angeles, 2005
'Curious Crystal of Unusual Purity', P.S.1 Contemporary Art Center, Long Island City, NY, 2004
'2004 Biennial Exhibition', Whitney Museum of American Art, New York, 2004
'Art at the Edge of the Law', The Aldrich Museum of Contemporary Art, Ridgefield, CT, 2001
'Twisted: Urban and Visionary Landscapes in Contemporary Painting', Stedelijk Van Abbemuseum, Eindhoven, The Netherlands, 2000
'Open Ends: MoMA 2000', The Museum of Modern Art, New York, 2000

Selected publications

Berry, Ian, *Fred Tomaselli*, Prestel Publishing, New York, 2009
Cappellazzo, Amy, *Fred Tomaselli: Ten Year Survey*, Palm Beach Institute of Contemporary Art, 2001
Volk, Gregory, *Fred Tomaselli*, James Cohan Gallery, New York, 2000
Tsai, Eugenie, *Gravity's Rainbow*, Whitney Museum of American Art at Philip Morris, New York, 1999

KEITH TYSON

Born in Ulverston, Cumbria, UK, 1969. Lives in Brighton, UK.

Turner Prize-winning artist Keith Tyson explores the complexities, mysteries and interconnectedness of human existence through the mediums of painting, drawing and installation. *The Block (Part of Seven Wonders of the World series)*, 2001–02, is one of a series of projects that explore scientific and philosophical concepts Tyson finds wondrous.

The centrepiece of *The Block* is literally a block of bronze. The twenty-three photographs in the installation depict a process of the block being melted and recast into many different sculptures, before being recast as a block. The first cast is of a model of the early universe, in the form of the Big Bang. Each recasting represents later phases in the cosmological understanding of the universe.

Tyson describes these transformations as the atomic memories of the bronze block. At its essence the work alludes to the fact that any given form is always temporary. Tyson uses this idea to reflect not just on cosmology, but also geological history, memory, culture, philosophy, and ultimately, art.

Selected solo exhibitions

'Keith Tyson', Blum & Poe, Los Angeles, 2009
'Fractal Dice', PaceWildenstein, New York, 2009
'Ten Years of Studio Wall Drawings 1997–2007', Haunch of Venison, London, 2007
'Large Field Array', De Pont Museum voor Hedendaagse Kunst, Tilburg, The Netherlands, 2007
'Keith Tyson', Kunsthalle Zürich, 2002
'From the Artmachine', Anthony Reynolds Gallery, London, 1996

Selected group exhibitions

'Wall Rockets: Contemporary Artists and Ed Ruscha', The Flag Art Foundation, New York; Albright-Knox Art Gallery, Buffalo, 2008–09
'Martian Museum of Terrestrial Art', Barbican Art Gallery, London, 2008
'Turner Prize Retrospective', Tate Britain, London, 2007
'Dionysiac', Centre Pompidou, Paris, 2005
'Century City: Art and Culture in the Modern Metropolis', Tate Modern, London, 2001
2nd Berlin Biennial, 2001
'Over the Edges', SMAK-Stedelijk Museum voor Actuele Kunst, Gent, Belgium, 2000

Selected publications

Holm, Michael Juul, *Large Field Array*, Louisiana Museum of Modern Art, Humlebæk, Denmark; De Pont Museum voor Hedendaagse Kunst, Tilburg, The Netherlands, 2006
Archer, Michael, *Geno Pheno*, PaceWildenstein, New York; Haunch of Venison, London, 2005
Macel, Christine, *Dionysiac*, Centre Pompidou, Paris, 2005
Tyson, Keith, *A Supercollider Notebook*, South London Gallery, London, 2002
Carey-Thomas, Lizzie and Katharine Stout, *Turner Prize 2002*, Tate Britain, London, 2002

THE POWER PLANT

The Power Plant Contemporary Art Gallery is a registered Canadian charitable organization supported by its members, sponsors, donors and funding bodies at all levels of government. We gratefully acknowledge the assistance of the Canada Council for the Arts, the Ontario Arts Council, Toronto Arts Council and Harbourfront Centre.

Canada Council for the Arts Conseil des Arts du Canada

ONTARIO ARTS COUNCIL
CONSEIL DES ARTS DE L'ONTARIO

torontoartscouncil
An arm's length body of the City of Toronto

Harbourfront centre

ACKNOWLEDGEMENTS

UNIVERSAL CODE: ART AND COSMOLOGY IN THE INFORMATION AGE DOCUMENTS THE 2009 group exhibition at The Power Plant. In realizing this ambitious exhibition and accompanying catalogue we are sincerely grateful for the support of Lead Donors Nancy McCain and Bill Morneau. We also thank Support Donors Elisa Nuyten and David Dime, Laura Rapp and Jay Smith, and Garnet and Evan Siddall, and Cultural Agency Supporter, the Consulat Général de France à Toronto. We would like to recognize the support of the Hal Jackman Foundation for providing free gallery admission for the duration of the exhibition. Thanks also to NOW Magazine, our Media Partner for the 'Universal Code' exhibition.

THE POWER PLANT IS INDEBTED TO THE ARTISTS WHO MADE THIS EXHIBITION POSSIBLE, and thanks the many lenders to 'Universal Code', including Albright-Knox Art Gallery, Buffalo; Thomas H. Bjarnason, Toronto; CNAP, Paris; Laura Rapp and Jay Smith, Toronto; Heidi L. Steiger, New York, and Norman and Norah Stone, San Francisco.

GENERAL THANKS

Albion Gallery, London; Claudia Altman-Siegel; Altman Siegel Gallery, San Francisco; Anthony Reynolds Gallery, London; Jade Awdry; Kristen Becker; Olivier Belot; Xenia Benivolski; Isabel Bernheimer; Pascale Berthier; Lara Blanchy; Michel Blancsube; Guillaume Bleret; Irene Bradbury; Jessica Bradley; Matthew Buckingham; Craig Burnett; Mircea Cantor; Elodie Cazes; Consulado General de México; Consulat Général de France à Toronto; Chantal Crousel; Donna Chu; Bellatrix Cochran-Hubert; David Zwirner, New York; Christine Davis; Benjamin Diaz; Donald Young Gallery, Chicago; Esther Schipper, Berlin; Dominique Fiat; Galleria Francesca Kaufmann, Milan; Galleria Franco Noero, Turin; Galerie Chantal Crousel, Paris; Galerie Dominique Fiat, Paris; Galerie Eva Presenhuber, Zurich; Galerie Micheline Szwajcer, Antwerp; Stevie Greco; Sophie Greig; Marie-Claire Groeninck; Haunch of Venison, London; Amelia Hinojosa; Antonia Hirsch; Julia Hölz; Ann Veronica Janssens; Jessica Bradley Art + Projects, Toronto; Jumex Collecion; Francesca Kaufmann; Katie Kehoe; Nina Köller; Jose Kuri; kurimanzutto, Mexico City; Matt Langton; Natalie Lazarus; Emily LeTourneau; Antoine Levi; Jed Lind; Luhring Augustine, New York; Sara MacDonald; Jochen Meyer; Meyer Riegger, Karlsruhe; Suzanne Modica; Gerardo Ochoa; Gabriel Orozco; Ales Ortuzar; Katherine Pill; Karen Pollack; Simona Pizzi; Eva Presenhuber; Anthony Reynolds; Thomas Riegger; Anjalika Sagar; Margherita Sassone; Joel Savary; Esther Schipper; Karen Schoellkopf; Hanna Schouwink; Corrado Tedeschi; Thea Westreich Art Advisory Services, New York; Jim Welter; Florian Weingrüll; White Cube, London; Jessica Witkin; Cerith Wyn Evans; Yvon Lambert, New York, Paris; Daniela Zarate

All Summer, All Free-Admission Sponsor:
Hal Jackman Foundation

All Summer, All Free-Media Partner:
NOW Magazine

PUBLISHED IN CONJUNCTION WITH THE EXHIBITION
UNIVERSAL CODE: ART AND COSMOLOGY IN THE INFORMATION AGE

THE POWER PLANT, TORONTO
12 JUNE—30 AUGUST 2009

THE POWER PLANT
231 QUEENS QUAY WEST
TORONTO, ON, CANADA M5J 2G8

PRINTED IN ICELAND

ISBN 978-1-894212-28-1

EDITOR/CURATOR: GREGORY BURKE
ASSISTANT CURATOR: JENNIFER MATOTEK
ESSAYS: GREGORY BURKE, JANINE MARCHESSAULT
TEXTS ON ARTISTS: GREGORY BURKE, JENNIFER MATOTEK
PROOF READER: DAN NACCARATO
DESIGN: ALISON HAHN, HAHN STUDIO
PRINTING & BINDERY: ODDI PRINTING
INSTALLATION PHOTOGRAPHY: RAFAEL GOLDCHAIN

Additional photo credits: pages 1, 10-11, 63, 116-120: courtesy White Cube, London; pages 4-5, 96-99: courtesy Galleria Franco Noero, Torino; pages 6-7, 40-43: photo Steve Payne; pages 8-9, 104-105: © Galerie Dominique Fiat, Paris. En cours d'acquisition collection CNAP, Paris, France; pages 12-13, 57-59: courtesy of the artists; pages 16, 76-81, 107-108: courtesy Galerie Micheline Szwajcer, Antwerp; page 25: courtesy Haunch of Venison. © Keith Tyson, 2009; page 26: courtesy Marian Goodman Gallery, New York; pages 44-45: courtesy Donald Young Gallery, Chicago; page 67: courtesy of the artist; pages 68-69: Courtesy Galleria Francesca Kaufmann, Milan; page 91: courtesy of the artist; page 92: courtesy of the artist; page 101: courtesy of the artist and kurimanzutto, Mexico City